ARTIFICIAL INTELLIGENCE FOR COMPUTATIONAL MODELING OF THE HEART

ARTIFICIAL INTELLIGENCE FOR COMPUTATIONAL MODELING OF THE HEART

Edited by

TOMMASO MANSI
TIZIANO PASSERINI
DORIN COMANICIU
Siemens Healthineers
Princeton, NJ, United States

ELSEVIER

ACADEMIC PRESS
An imprint of Elsevier

Academic Press is an imprint of Elsevier
125 London Wall, London EC2Y 5AS, United Kingdom
525 B Street, Suite 1650, San Diego, CA 92101, United States
50 Hampshire Street, 5th Floor, Cambridge, MA 02139, United States
The Boulevard, Langford Lane, Kidlington, Oxford OX5 1GB, United Kingdom

Library of Congress Cataloging-in-Publication Data
A catalog record for this book is available from the Library of Congress

British Library Cataloguing-in-Publication Data
A catalogue record for this book is available from the British Library

ISBN: 978-0-12-817594-1

For information on all Academic Press publications
visit our website at https://www.elsevier.com/books-and-journals

Publisher: Mara Conner
Acquisition Editor: Chris Katsaropoulos
Editorial Project Manager: Isabella C. Silva
Production Project Manager: Surya Narayanan Jayachandran
Designer: Miles Hitchen

Typeset by VTeX

Contents

List of figures

List of contributors

Dorin Comaniciu, Bogdan Georgescu, Florin C. Ghesu, Sasa Grbic, Lucian Mihai Itu, Tommaso Mansi, Viorel Mihalef, Dominik Neumann, Tiziano Passerini, Saikiran Rapaka, Puneet Sharma, Yue Zhang
Siemens Healthineers, Princeton, NJ, United States

Helene Houle
Siemens Healthineers, Ultrasound Division, Mountain View, CA, United States

Felix Meister
Siemens Healthineers, Princeton, NJ, United States
Friedrich-Alexander University Erlangen-Nuremberg, Erlangen, Germany

Cosmin Nita, Andrei Puiu
Siemens SRL, Image Fusion and Analytics, Brasov, Romania

Foreword

The clinical cardiology community is witnessing an exciting disruption towards precision medicine, fueled by two revolutionary quantitative science approaches: computational modeling and artificial intelligence (AI).

Precision medicine is becoming the dominant paradigm in the contemporary practice of medicine. Challenging the traditional population-based model, it attempts to tailor medical treatment to the individual characteristics of each patient, quantified through genomic, proteomic, metabolomic and physiomic features. Examples of precision medicine already in use today include genetic testing for inherited arrhythmia syndrome (e.g. KVLQT1, HERG, SCN5A) to prevent sudden cardiac death, familial hypercholesterolemia to prevent coronary artery disease, pharmacogenomics to assess individualized response to critical drugs (e.g. CYP2C19, SLCO1B1, VKORC1), and measurement of cardiac troponin I (cTnI) to risk-stratify patients following acute coronary syndrome.

Medical imaging is playing a growing role in precision medicine. Over the past decade, advanced cardiovascular imaging such as cardiac computed tomography (CT), magnetic resonance (MR) and positron emission tomography (PET) has succeeded in providing patient-specific anatomical and physiological information that allows individualized, minimally invasive treatment in clinical cardiology. For example, myocardial scar imaging by MR provides individualized information regarding the risk of fatal arrhythmia and the potential target of interventional therapy to prevent arrhythmic events. From a mathematical point of view, imaging provides a detailed mapping of cardiac anatomy and substrate, represented as a time-varying scalar field of signal intensity (Fig. 0.1 *Imaging*). Some physiological information such as blood flow velocity, myocardial blood flow and finite deformation of the cardiac tissue can also be measured, but it is typically limited in spatial and temporal resolution.

While imaging is excellent at capturing heart shape, substrate and kinematics, understanding and quantifying how it functions is still an open challenge. To that end, researchers are looking for ways to combine medical imaging and computational modeling to provide detailed physiological information based on physical principles (Fig. 0.1 *Computational Modeling*). Thanks to recent technological breakthroughs, image-based, patient-specific computational modeling is becoming a powerful tool to augment the

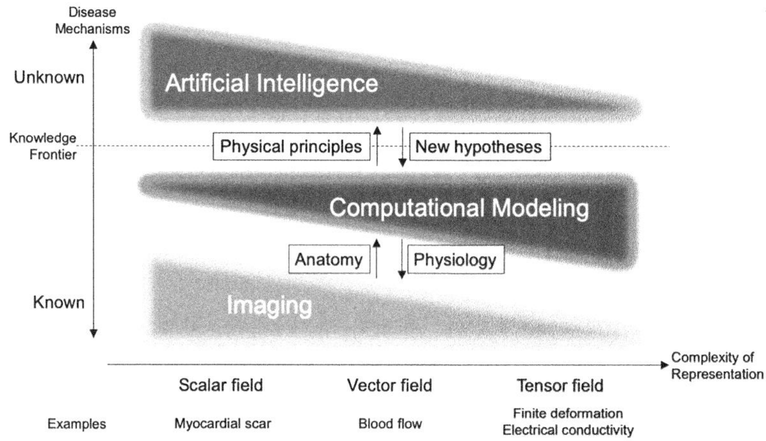

Figure 0.1. How AI and computational modeling can push the knowledge frontier for precision medicine.

visible data with physiological knowledge. For instance, fractional flow reserve (FFR) can now be estimated from CT to identify physiologically critical stenoses in coronary arteries. MR-based methods for target identification in complex arrhythmia such as atrial fibrillation and ventricular tachycardia are becoming a reality. Patient-specific modeling has therefore the potential to augment imaging with quantitative features of heart physiology and its underlying biophysics, mathematically represented as time-varying vector or tensor fields.

With the increasing availability of very large medical database comprising imaging, reports, patient record, etc., artificial intelligence (AI) has the potential to disrupt healthcare as we know it, similar to what we are witnessing in other industries. AI consists in learning complex models directly from large databases. Deep learning, in particular, unlocks the discovery of data features and patterns that allow higher levels of predictive performance compared to traditional machine learning approaches. AI they already allowed unprecedented success in various applications, such as cardiac motion analysis for survival prediction and human-level arrhythmia detection in electrocardiogram, to cite just a few examples. Nevertheless, AI algorithms, especially deep learning, are hard to examine after the fact to understand specifically how and why a decision has been made. This lack of explainability of AI hampers our ability to improve mechanistic understanding of the disease and develop effective therapies. For example, AI can pick up the features of low cardiac function in 12-lead electrocardiogram that were previously unrecognized, but it remains unknown what specific features in 12-lead electrocardiogram represent cardiac function, and why they are related to heart failure. Physiological modeling can complement AI-based analysis by on the one hand enabling the exploration of new hypotheses about unknown mechanisms; on the other hand, informing AI of physical principles of known mechanisms to potentially address the lack of explainability (Fig. 0.1 *Artificial Intelligence*). Together, AI and computational modeling can push the knowledge frontier further for understanding disease mechanisms in a patient-specific manner and accelerate the evolution of precision medicine.

This book is a perfect introduction to the interdisciplinary science of computational modeling and AI, and how those disciplines could help disrupt clinical cardiology practice. Drs. Mansi, Passerini and Comaniciu put together the top scientists in Siemens Healthineers at the forefront of this technological revolution. They are among the few people who can understand and speak the language of both quantitative and clinical sciences. This book is born out of integrated expertise in cutting edge technologies as well as deep understanding of challenges and limitations of the current standard of care. This book is an exciting contribution to the field, and it is recommended to anyone interested in using new and bold approaches to save people's lives and shape the health care of the future.

Associate Professor of Medicine, MD, PhD
Hiroshi Ashikaga
Cardiac Arrhythmia Service, Division of Cardiology
Johns Hopkins University School of Medicine
Baltimore, MD, United States
September, 2019
Lund, Sweden

Preface

What if we could create a digital representation of a person's heart, that functions and beats like the real organ? What if the virtual heart could react to physiological stresses, diseases, and interventions as if real? This concept, known as digital twin, is well established in the world of manufacturing and industry [1]. Many engineering artifacts, from airplane to cars to highly complex wind turbines, among others, are first developed numerically. The resulting digital twins are nowadays commonly used to predict failure, plan maintenance strategies, or even guide operations in a changing environment [2]. Digital twins are also used to train robotic systems powered by artificial intelligence (AI), before deploying them in the field [3]. With the increasing cost of physical testing, due to the increasing complexity of the manufactured systems in particular, it is not a surprise that digital twins are becoming a key engineering tool in manufactory and industry.

What if that same concept would be possible in healthcare? One could use a digital twin of a person to predict disease, derive personalized risk scores and allow patients to manage their health proactively, enable physicians to optimize therapies for specific patients, and much more. Digital twins could become a key component of the new wave of precision medicine [4–6].

Estimating the digital twin of a physical artifact is not a simple task [1]. First, it requires designing and implementing accurate models of the physics underlying the function of the artifact. In industry, the artifact is created from scratch and directly designed using computer assisted design software. Its physics, for instance mechatronics, is therefore known by design. Second, the computational model of the artifact needs to be individualized. In other words, the parameters of the model need to be adjusted such that the digital twin replicates how the artifact of interest works along with its state, including potential defects in the object, the effects of previous events, and physical properties. The individualization task needs to be performed continuously, as data from sensors flow, to keep the digital twin up to date. Finally, interactions with the environment need to be accounted for, such as the effects of natural elements, human interactions, or simply long term wear of the artifact materials. This step is crucial if one wants to use the digital twin for predictive maintenance, error recovery or simply operational optimization [2].

Creating digital twins of biological systems can be daunting. Contrary to industrial artifacts, the underlying biophysics are still not fully understood. The high fidelity physiological models that have been so far are highly non-linear, with multiple feedback loops and controls, spanning spatio-temporal scales going from the molecules to the body, from nano seconds to many years [7]. Furthermore, biological data can require invasive or destructive measurement procedures, they may be noisy, or too sparse and incomplete. Despite these challenges, tremendous progress towards patient-specific modeling of organ and system function has been made in the past decades, with initial clinical proof of concepts and validation already happening [5,6,8,9]. The heart is one of the organs that has attracted significant interest from biophysiologists, mathematicians and modelers. Researchers first used mathematical models to understand heart function and the underlying physiology, progressively moving towards clinical applications like the characterization of heart failure. Evolution of technology, computational power and healthcare sensing technologies (e.g. imaging scanner, electrophysiology sensors, etc.) has been pivotal towards this progression [4].

In parallel, we are witnessing a new wave of artificial intelligence developments [10]. Fueled by the exponential increase of available digital data and computational power, new methods based on deep neural networks are being developed, disrupting a wide range of fields, from computer vision to natural language processing, and now including computational biology, physics and healthcare applications. Provided a large amount of representative data is available, deep learning algorithms learn with unprecedented effectiveness the feature patterns that are most relevant for a specific task (e.g. classification or prediction), reaching unparalleled levels of performance [11].

Motivated by these two trends, the rise of digital twins and artificial intelligence, this book presents opportunities arising from the combination of both fields. We believe that the performance achieved by AI systems, informed by biophysical knowledge and models, could provide a new generation of tools for improved health management, with potential impact on increased favorable treatment outcomes and clinical workflow optimization. While traditional computational models of the heart are complex to use, requiring specific expertise to build and manipulate them, AI could foster a new way of integrating available data and physiological models. Specific modeling components could be built automatically, such as the anatomical model that is created from multi-modality imaging data. Modeling tasks could be made dramatically more computationally efficient, such as the estimation of individualized model parameters or the prediction of quantities of clinical interest. Additional and more powerful feature patterns could be extracted from the data, when the output of physiological modeling is considered as additional information. This could finally promote the use of digital twins of human organs and systems where they are most needed: at the best side.

Computational modeling of the heart: from physiology understanding to patient-specific simulations

Finding mathematical laws that explain physiological systems has been of scientific interest for centuries. Leonardo Da Vinci's Vitruvian man provides an outstanding example of the quest for mathematical laws of human proportions and beauty. However, it is commonly acknowledged that mathematical models of cardiac cells and tissue were first proposed in the 1950's, as the understanding of biology increased in the scientific community. In particular, Hodgin and Huxley's work has been seminal in deriving what is known to be the first model of the cellular action potential [12]. Since then, tremendous progress has been made in modeling heart biophysiology, from the cardiac action potential [13], sarcomeres function [14], excitation-contraction coupling [15], metabolism [16], blood perfusion and then going up to the tissue level, organ level and finally circulatory and human body levels [7]. Multiple models have been proposed and made available in public repositories (e.g. CellML [17]), with varying degrees of complexity and fidelity with respect to wet lab experiments. However, it still remains an art to identify which model to apply, what simplifications to make and which parameters to adjust to simulate the biophysics under investigation.

Supported by the growing computational power along with increasing amount of clinical data, researchers have started to investigate methods to perform patient-specific simulations [5,6, 18]. In this endeavor, a new set of challenges arises. How detailed should the model be to capture the disease of the patient? What data is available in a clinical setup to estimate a useful patient-specific model? If some parameters are not clinically measurable, what should their value be? What is a good validation strategy? How should these models be integrated into the clinical workflow? These are only a small sample of the questions that need to be answered for a clinically useful virtual heart. Recently published works are suggesting promising results, like for instance non-invasive characterization of cardiac function in heart failure [6], planning and guidance of ablation therapy for cardiac arrhythmias [6,8,9,19–21], valve surgery planning [22,23] or congenital heart disease treatments [24,25]. Yet, physiological models are still built and used by modelers and scientists, outside of the clinical environment. They are also still computationally demanding and complex to operate, making their large scale, prospective clinical validation difficult. Finally, the effects of data noise, parameter uncertainty and model assumptions still need to be properly assessed [26,27].

The new age of artificial intelligence

The concept of intelligent machines can be traced back to the origin of human made machines. However, the term "artificial intelligence" (AI) was first introduced by John McCarthy, one of the founding father of AI, during the Dartmouth conference in 1956 [28]. Since then, the AI community has gone through several waves of hype and winter periods. Nowadays, AI is a core technology in many fields, powered by what is called deep learning. Deep learning methods rely on artificial neural networks (ANN) with a very large number of layers. While ANNs were first proposed in 1943 [29], their power could be unlocked only recently thanks to the availability of large amount of data along with the tremendous computational power offered by high-performance computing infrastructures. In computer vision for instance, one could consider the 2012 AlexNet convolutional neural network [30] as one of the main trigger of modern deep learning, along with the ImageNet database (2009), which facilitated an exponential development in computer vision technologies.

AI did not wait long to penetrate medical imaging. Machine learning methods were already explored in the 2000's for organ detection, contouring and tracking [31,32]. As in computer vision deep learning has provided a significant boost in performance for each of these core image processing tasks, but also image reconstruction [33] or even scanner automation. As an example, an AI algorithm has been shown to be able to estimate an avatar of a patient laying on the table of a computed tomography scanner. The avatar is then used by the system to perform automatic isocentering or scan range estimation to image a specific organ [34, 35]. As large databases with outcome data are becoming available, researchers are investigating how AI could help estimating new risk scores with significantly better sensitivity and specificity than the state of the art. For instance, in [36], the authors show that an AI algorithm can calculate a risk score for radiotherapy response that is 45% more discriminative than tumor size only, the current criterion for clinical decision making. With AI, workflows are poised to become more and more automatic, while new decision support tools will appear. For these reasons, healthcare is believed to be one industry to be significantly disrupted by AI.

Computational modeling meets statistical learning

Interestingly, both fields of computational modeling and AI started almost at the same time, and both benefited from the recent access to data and computational power to unlock their potential in many applications. A recent scientific direction consists in exploring ways to augment computational models with statistical methods, and conversely. A first objective is model reduction for fast but accurate simulations. The idea is to use a data-driven approach, typically manifold learning, to reduce complex models into a reduced set of equations that is easier to tract while capturing the overall trend of the physics under considerations. In [37] for instance, the authors applied dimensionality reduction to complex ionic models of sarcomeres to simplify and speed up their computational solution. The expressive power of deep learning however could also allow to predict the dynamics of complex physical systems directly from observable input features. For instance in [38], the authors show that a neural network can simulate the universe evolution under gravity, a complex N-body system, as accurately as the most sophisticated simulator while being several orders of magnitude faster. Not only, the deep neural network could also predict universe evolution under conditions not observed during training. Similar approaches are being investigated for computational fluid dynamics [39] and biomechanics [40], mostly in the field of computer graphics, to develop fast and accurate solvers.

Another fascinating area is the use of data and computational models to train neural network systems that predict the occurrence of scenarios of interest. In [41], the authors train a model to predict the next occurrence of earthquakes in a region, given sparse and noisy seismographic data. Interestingly, the authors found that the output of the neural network could be explained in a physical sense. This highlights an interesting trend, where data-driven models informed by training samples produced by computational models can help generate new hypotheses, that could then be further analyzed through computational modeling, and even be used to augment these models.

In the same spirit, this book describes how computational modeling and AI can be integrated for the study of human physiology and pathology. The reader will discover methods and approaches that combine the best of both sciences, for fast, accurate and personalized digital twins of the heart.

Book organization

The book is organized into two parts. First, general cardiac physiology, modeling, and implementation concepts are introduced. The objective is to provide the reader with the necessary background that will then be useful for the remainder of the book. The second part describes how machine learning and artificial intelligence can facilitate patient-specific modeling, ranging from automatic anatomical modeling to AI accelerated models to robust parameter estimation. Finally, the book concludes with clinically relevant examples to illustrates how such technologies could be applied to real-world applications.

Part 1: Modeling the beating heart: approaches and implementation

Chapter 1: Multi-scale models of the heart for patient-specific simulations. The first chapter of the book introduces the main concepts and models of heart function. Physiological mechanisms and related models are introduced and discussed. The major physiological systems are described: anatomy, electrophysiology, hemodynamics and biomechanics. Methods used for parameter estimation from clinical data are also introduced.

Chapter 2: Implementation of a patient-specific cardiac model. The second chapter deep dives in one exemplary model and describes implementation strategies for each physiological systems. More precisely, lattice-Boltzmann methods are presented for computational fluid dynamics and electrophysiology. Graph-based approaches for real-time cardiac electrophysiology simulations are also detailed. A computationally efficient finite element method for modeling soft tissue deformations, based on the Total Lagrangian Explicit Dynamics framework, is introduced for the computation of bi-ventricular electromechanics. Finally, the chapter closes with an integration strategy to efficiently perform fluid-structure interaction simulations.

Part 2: Artificial intelligence methods for cardiac modeling

Chapter 3: Learning cardiac anatomy. The first step in patient-specific modeling is the fast and robust parsing of the cardiovascular anatomy from medical images. This task usually involves detecting, segmenting and tracking anatomical structures or pathologies in the human heart. Current approaches are based on machine learning methods that scan the image and calculate a probability for each voxel to belong to the anatomy of interest (e.g. myocardium). As a result, these methods suffer from inherent limitations related to the efficiency in scanning high-dimensional parametric spaces and the learning of representative features for describing the image content. This chapter introduces new techniques for cardiac image parsing and structure tracking. First, the marginal space learning framework is introduced, including the original version of the system that relies on handcrafted steerable features, as well as the modern redesign of the framework based on the latest automatic feature learning technology using deep learning. We then introduce the concept of intelligent image parsing based on deep reinforcement learning and scale-space theory, which overcomes the inherent limitations of the previous methods. Third, a modern deep image-to-image neural network architecture for whole heart isolation is presented. Finally, a review of cardiac structure tracking approaches based on convolutional neural networks and recurrent neural networks is provided.

Chapter 4: Data-driven reduction of cardiac models. A common challenge in physiological modeling is to find the right trade-off between model fidelity, and hence complexity, and computational efficiency. Digital twins can be integrated into clinical workflows only if the data processing is time efficient and user-friendly. In other words, all complexity needs to be hidden from the user. In the ideal case, detailed and individualized physics phenomena should be computed in real-time, to allow the user to interact with the virtual heart. Unfortunately, current approaches to real-time simulations come with significant model simplifications, which may not accurately reflect the complex physiology any more. Artificial intelligence could open new ways of simplifying models, while retaining their descriptive power. This chapter describes data-driven approaches for model reduction, with the aim of retaining the ability of the model to describe complex physical phenomena, while at the same time drastically reducing the computational complexity. Three use cases are presented, covering cardiac electrophysiology, hemodynamics and biomechanics.

Chapter 5: Machine learning methods for robust parameter estimation. The second challenge for accurate patient-specific simulations is data assimilation. In particular, models need to be fitted to patient data in order to reproduce the observed physiology. In clinical practice, data is often sparse, noisy and incomplete. For instance, 12-lead electrocardiograms only provide an indirect view of cardiac electrophysiology. The inverse problem of parameter estimation is also often ill-posed. Traditional approaches rely on optimization techniques with ad hoc regularizers or optimization strategies. This chapter explores how machine learning methods can be used to solve the inverse problem. A first approach consists in learning a regression model that maps clinical measurements to model parameters directly using a large database of synthetic data. The method is illustrated for the case of cardiac electrophysiology. A second approach builds upon reinforcement learning techniques to learn the optimization process itself, leading to a potentially more efficient and more robust algorithm than black box optimization methods.

Chapter 6: Additional clinical applications. The last chapter of the book presents three additional clinical applications of the methods presented in the previous chapters. First, a multi-scale, multi-physics cardiac model for cardiac resynchronization therapy (CRT) is presented. CRT is an effective treatment of heart failure. It consists in placing an advanced pacemaker that stimulates both right and left ventricles to resynchronize the failing heart. Unfortunately, the therapy response rate is still low. Methods that could guide the electrophysiologist on where to implant the stimulating electrodes and how to program them are needed. In the presented work, the model is first built from pre-operative, standard-of-care data. It is then used to simulate various pacing protocols. Prediction results are compared with observed changes in cardiac electrophysiology on ten patients illustrating the use of the virtual heart to visualize the effects of the treatment before delivering it. The second application relates to aortic coarctation (CoA), a congenital disease that requires surgical treatment. The decision to treat is based on the pressure drop across the coarctation. This is assessed using an invasive procedure, with pressure catheters. We present a method that allows the computation of the pressure drop directly from images by learning a model from data generated using computational fluid dynamics. Finally, the third application addresses the entire cardiovascular system, based on a lumped parameter model (LPM) of whole body circulation. A similar approach to the one employed for the CoA use case is considered, relying on a purely synthetic training database. Both time independent, e.g. systemic resistance and compliance, and time dependent quantities, e.g. ventricular pressure, are predicted. We show that the performance of the AI models is statistically similar to that of the LPM, while the inference time is reduced to milliseconds.

Outlook

As we wrote this book, our goal was to introduce the reader to the fascinating field of AI-augmented biophysical modeling. We are convinced that combining these two worlds will enable the development of new, disruptive solutions to help physicians decide what is best for their patient, guide them towards the optimal therapy: in brief, to enable a more precise medicine. The field is moving very fast, and this book is only a snapshot of what the community is currently researching and working on. However, we hope it will inspire the reader to explore hybrid approaches, where both physical knowledge and AI enrich each other. We cannot wait for the next developments that will happen in the years to come. What if we could have our own digital twin?

Disclaimer: The concepts and information presented in this book are based on research results that are not commercially available.

Tommaso Mansi, Tiziano Passerini, Dorin Comaniciu

List of abbreviations

2D 2 dimensional
3D 3 dimensional
AFib Atrial fibrillation
AHA American Heart Association
AI Artificial intelligence
AV Atrio-ventricular
BEM Boundary element method
BSP Body surface potential
BSPM Body surface potential mapping
CFD Computational fluid dynamics
CICR Calcium-induced calcium release
CPU Central processing unit
CoA Coarctation of the aorta
CRT Cardiac resynchronization therapy
CT Computed tomography
DCM Dilated cardiomyopathy
DL Deep learning
DRL Deep reinforcement learning
ECG Electrocardiogram
EM Electromechanics
EP Electrophysiology
FDM Finite difference method
FEM Finite element method
FSI Fluid-structure interaction
GPU Graphical processing unit
Graph-EP Graph-based solver for cardiac electrophysiology
HD Hemodynamics
HO Holzapfel–Ogden
LA Left atrium
LBM Lattice-Boltzmann method
LBM-EP Lattice-Boltzmann method for cardiac electrophysiology
LPM Lumped parameters model
LV Left ventricle
M-S Mitchell and Schaeffer model
ML Machine learning
MRI Magnetic resonance imaging
MSDL Marginal space deep learning
MSL Marginal space learning
MV Mitral valve
ODE Ordinary differential equation
PDE Partial differential equation
PV Pulmonary valve
PVJ Purkinje fiber – ventricular muscle junction
RA Right atrium
RV Right ventricle
SPK Second Piola Kirchhoff stress tensor

SR Sarcoplasmic reticulum
TLED Total Lagrangian Explicit Dynamics
TV Tricuspid valve
VT Ventricular tachycardia

1

Modeling of the beating heart: approaches and implementation

Multi-scale models of the heart for patient-specific simulations

Viorel Mihalef, Tiziano Passerini, Tommaso Mansi
Siemens Healthineers, Princeton, NJ, United States

Modeling heart function consists in translating the main mechanisms that govern heart function into mathematical laws and computational algorithms. Multi-scale models describe the interactions at different time scales (from nanoseconds to minutes to years), spatial scales (from nanometers to centimeters), and functional scales (from molecular pathways to circulatory system). Such models can therefore pave the way to predictive medicine by computing advanced measurements, planning therapies in-silico before the intervention, evaluating the effects of molecular changes on the global cardiac function. Scientific and technological development of such models are enabling comprehensive and detailed understanding of heart physiology. At the same time, clinical translation requires a clear clinical focus on selecting the models, controlling the underlying assumptions and simplifications, and ensuring their usefulness to address real-world clinical problems while being integrated into the day-to-day workflow. This chapter introduces the typical components of a multi-scale heart model that includes personalized anatomy, electrophysiology, biomechanics, valves, hemodynamics, and common approaches to parameter estimation from routinely available clinical data.

1.1 Models of cardiac anatomy

Let us first review the major anatomical elements to consider for cardiac modeling. The reader is invited to refer to specialized literature such as [42] for a more detailed description of cardiac anatomy and function. The very first step in modeling heart function consists in defining a realistic model of its anatomy, which includes shape, tissue types (also called substrate) and tissue micro-architecture. At a high level, the heart can be described as a system of four blood-filled chambers separated by valves (Fig. 1.1). Functionally and anatomically the heart can be divided into the atria

Artificial Intelligence for Computational Modeling of the Heart
https://doi.org/10.1016/B978-0-12-817594-1.00011-5

and ventricles, as well as a left (systemic) and right (pulmonary) side. The oxygen-poor blood enters first the right atrium, then the right ventricle through the tricuspid valve. The right ventricle then ejects the blood into the lungs through the pulmonary valve. The blood gets oxygenated by the lungs and then goes back to the heart in the left atrium chamber, from where it reaches the left ventricle through the mitral valve. Finally, the left ventricle ejects the blood through the aortic valve into the circulatory system. The muscle, called myocardium, of the atria is much thinner than ventricular myocardium as they only need to transfer the blood to the ventricles. The ventricles have to provide a more significant amount of work to push blood in the peripheral circulation. The heart is enclosed into a relatively stiff bag called pericardium, which is fixed in the thoracic cavity. A thin layer of pericardial fluid between the pericardium and the external layer of the myocardium (called epicardium) ensures frictionless motion of the heart inside the pericardial bag.

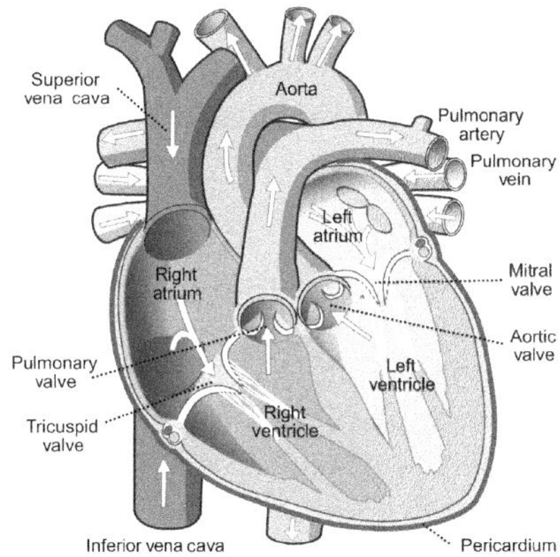

Figure 1.1. Diagram of heart anatomy and blood flow. (*Source: Wikipedia.*)

Depending on the application, various anatomical models have been used by researchers. Regarding the ventricles, the most common model for physiology and disease understanding is surely based on analytical formulations. The left ventricle is modeled using a prolate spheroidal model [43,44], which approximates relatively well its healthy shape. In these models, the other chambers are ignored. Generic bi-ventricular models have also been created to investigate inter-ventricular interactions and obtain more realistic results [45] (Fig. 1.2, left panel). As detailed 3D imaging became possible, researchers quickly moved to more realistic, and sometimes also pathological geometries. First, cardiac anatomies extracted from 3D scans of animals were used [46], and

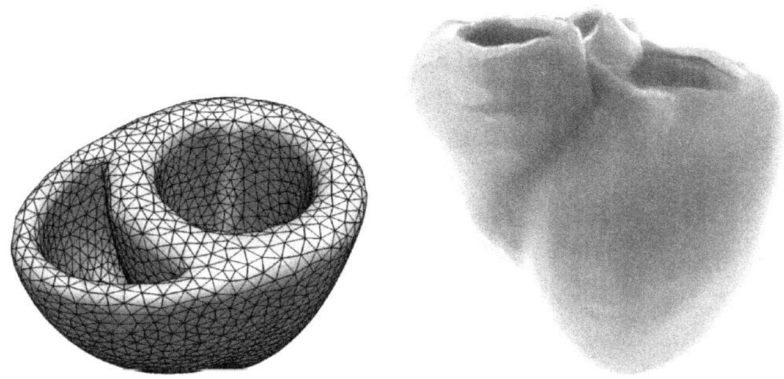

Figure 1.2. Examples of heart geometries. *Left panel*: analytical, prolate spheroidal model of the two ventricles. *Right panel*: patient-specific model estimated from medical images.

recently geometries from medical images of patients have become standard [6,9,47] (Fig. 1.2, right panel). Chapter 2 describes how to estimate such a patient-specific geometric model.

Analytical models of the atria are much more difficult to design, due to their complex shape and multiple vessel insertions. As a result, researchers often used animal-based atlases in their studies. Patient-specific models became available only recently, with the diffusion of high-resolution 3-dimensional (3D) computed tomography (CT) scans and magnetic resonance imaging (MRI).

Valves have also been modeled with various levels of fidelity. Analytical models have first been designed to investigate their physiology in various conditions [48]. With the development of 3D transesophageal echocardiography (TEE), valves can be now imaged in-vivo, allowing patient-specific modeling of the leaflets and surrounding structures [32]. Finally, the pericardium, which is hardly visible in images, is often approximated as a shell whose shape is given by the epicardium [6].

Once the shape of the structure of interest is modeled, two other components need to be considered: the micro-architecture (i.e. cellular organization) and substrate of the myocardium. The myocardium is organized in fiber bundles, themselves organized in sheets. The orientation of the fibers varies from apex to base, and from epicardium to endocardium (the inner layer of the myocardium). Myocardium fibers play a crucial role in cardiac function. In particular, they affect biomechanical and electrical properties. Integrating them into the model is therefore essential. The distribution of the fibers across the myocardium has been widely studied on ex-vivo hearts using histology or diffusion tensor MR imaging (DTI) [49–51]. Because measuring them in-vivo is still an open challenge [52], reference atlases of fiber orientations are being computed based on dogs hearts [53] or human hearts [54]. Alternatively, rule-based models have been developed, based on ex-vivo studies [49,55–57], and still widely used (see Fig. 1.3).

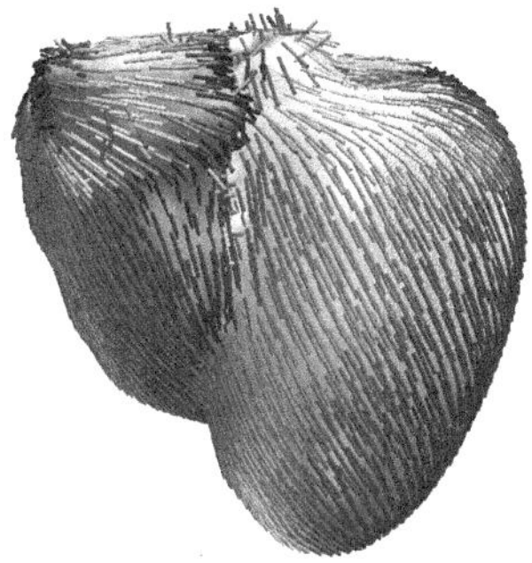

Figure 1.3. Example of heart fiber model computed using rule-based approach on a patient-specific heart anatomy.

The last element for a comprehensive anatomical model is substrate. Diseases can significantly affect the cellular structure of the myocardium tissue [42]. Fibrotic tissue can accumulate between the myocytes (cells that form the myocardium and are responsible for the active contraction), which makes the overall tissue stiffer and electrically impaired. The myocardium could also suffer from infarction due to the lack of blood perfusion, as a consequence of coronary artery disease. The infarcted area is physiologically inactive, namely not electrically conductive nor contracting. This can impact the function of neighboring tissue as well as the heart overall. Both fibrosis and infarct scar are therefore crucial to capture for precise modeling in diseased hearts. MRI is the modality of choice to visualize scars and fibrosis. One can use for instance delayed-enhancement MRI, with gadolinium injection, to image fibrosis and scar (see Fig. 1.4). The region is segmented by an expert or using image processing algorithms, and included into the anatomical model. Chapter 2 gives more details on how this is done in practice.

1.2 Electrophysiology modeling

Heart function is determined by the coordinated, rhythmic contraction of the chambers, displacing blood to the body. Such coordination requires an effective activation system that ensures synchronized contraction. This functionality is achieved through the cardiac electrical system, and is called electrophysiology (EP).

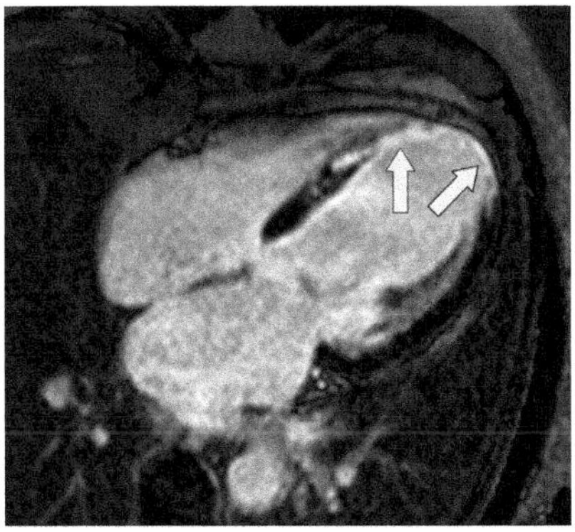

Figure 1.4. Example of MRI with delayed enhancement of gadolinium highlighting a transmural scar (yellow arrows (light gray in print version)) in the septum and apex of the heart. (*Source: Wikipedia.*)

The myocytes are capable of dynamically altering the electrical potential across their cellular membrane, thus generating electrical signals. The dynamics of these signals is determined by the balance of concentration of ionic species in the intra- and extracellular environments. The cellular membrane is permeable to these ionic solutions through so called ion channels. These channels can selectively open or close depending on metabolic conditions as well as environmental stimuli. In normal physiological conditions, specialized myocardial cells in the sino-atrial node, the natural pacemaker located at the junction of the superior vena cava with the right atrium, spontaneously generate periodic electrical signals by regulating the balance of ionic species across their membrane. These electrical signals affect the behavior of neighboring cells: myocytes react to changes in the extracellular electrical potential by adapting the permeability of the cellular membrane through the activation of specialized ionic channels. This effectively allows the electrical signal to propagate in the cardiac tissue, as the cells progressively adapt their trans-membrane electrical potential in response to the activation of neighboring tissue. The altered trans-membrane potential also triggers additional changes to the membrane permeability, which eventually restore the original equilibrium of ionic species in the intra- and extra-cellular environments, thus setting the stage for a new activation cycle.

The propagation of the electrical signal is driven by the electrical conduction system of the heart, specialized tissue responsible for transmitting the electrical signal from the sino-atrial node to the rest of the myocardium in a controlled way (Fig. 1.5). Myocytes

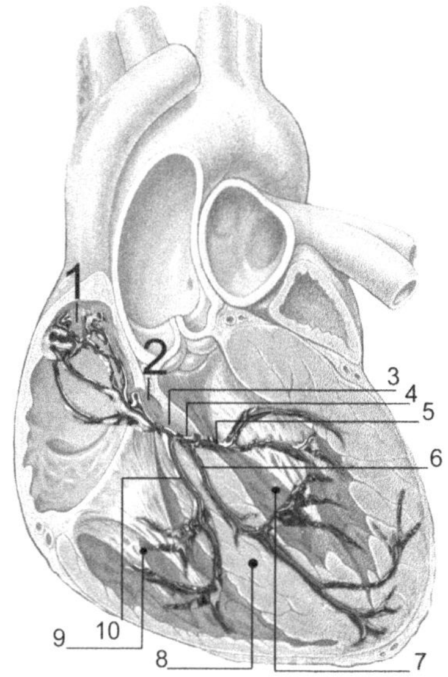

Figure 1.5. Schematic depiction of the electrical conduction system of the heart. 1. Sinoatrial node, 2. Atrioventricular node, 3. Bundle of His, 4. Left bundle branch, 5. Left posterior fascicle, 6. Left anterior fascicle, 7. Left ventricle, 8. Ventricular septum, 9. Right ventricle, 10. Right bundle branch. (*Source: Wikipedia.*)

belonging to the electrical conduction system are electrically insulated by fibrous sheathing from the non-specialized myocardium, with the exception of junction points that allow the controlled delivery of the electrical stimulus along the path of the conduction. It should be noted that non-specialized myocytes also conduct the transmembrane potential, although at a slower speed. From the sino-atrial node and through dedicated pathways in the right atrium, the signal reaches the atrioventricular (AV) node located in the lower part of the right atrium on the fibrous atrioventricular ring. Fibrous tissue separates the musculature of atria and ventricles, the only electrical connection being the muscular bundle of His departing from the AV node. After crossing the atrioventricular junction, the bundle of His splits (usually in two branches) at the summit of the ventricular septum. The left bundle branch is a series of fascicles spreading over the septum of the left ventricle and connecting with ordinary myocardial fibers. The right bundle branch tends to remain a bundle until it reaches the anterior papillary muscle of the right ventricle, where it splits in fascicles spreading over the right ventricle myocardium. The tree-like terminal branchings of the left and right bundles are called Purkinje fibers and extend subendocardially up to one third or one half of the ventricular thickness [58]. The electrical pulse propagates along the bundle of His and its terminal fibers with a speed that is about four times larger than in the myocardial tissue [59]. As a result, the depolarization of the ventricle is triggered in the endocardium and then reaches the epicardium.

Detailed studies of the electrical conduction system have been based on histology as well as invasive measurements of electrical activation [60]. Such studies shed light on the role and function of the specialized myocardial tissues. For instance, early studies suggested that the network of Purkinje fibers is densely diffused in the subendocardium, and that it connects to the ventricular muscle in discrete regions called Purkinje fiber-ventricular muscle junctions (PVJ) [61].

The coordinated electrical activity of the myocardium induces a dynamic change of the electrical potential in the tissues and organs surrounding the heart. This produces a measurable electrical signal on the surface of the body, which can be captured by electrocardiography. This is the most accessible measurement of the electrical activity of the heart, and has been used as a valuable tool to understand the healthy electrophysiology as well as detect pathological alterations [62]. Invasive measurement of electrical activity on the endocardial surface additionally allows the assessment of the role of local substrate properties and abnormalities (such as the presence of scarred or fibrotic myocardial tissue) on the activation pattern. For this reason, it has become a crucial tool for instance in the planning and delivery of ablation therapy for cardiac arrhythmias [63].

Physiological modeling of cardiac electrophysiology aims at identifying and understanding the underlying biophysical phenomena, with the ultimate goals of allowing risk stratification of patient populations based on estimated model parameters; and designing interventional strategies for the optimal treatment of electrical heart diseases such as cardiac arrhythmias. These have been topics of interest in the research community and have sparked the development of various modeling approaches. Seminal work by Hodgkin and Huxley introduced a modeling framework to study cellular excitability, paving the way for computational modeling of electrophysiology at the cellular and tissue level [12,64–66].

The following sections provide a brief overview of the most common modeling approaches, including cellular, tissue and body level models.

1.2.1 Cellular electrophysiology

Historically, *phenomenological models* were the first models to be developed, derived from experimental observations on nerves or myocardium tissues [64,65]. These models reproduce the observed shape of the action potential and how it changes under the effect of external conditions, without focusing on the underlying ionic phenomena. Hence, they provide a simplified representation of the dynamics of the cell, which limits their ability to capture changes in the electrophysiology due to microscopic effects, for instance as a result of the action of pharmacological agents. On the other hand, these models are typically controlled by a limited number of model parameters, which makes them computationally efficient. Moreover, the parameters are typically directly related to the shape of the action potential or ECG measurements, which enables effective parameter estimation strategies from clinically available data.

Biophysical models simulate the ionic interactions across the cell membrane and the biological phenomena underlying ion channels [13,67–69]. They are largely based on direct measurements of electrical signals in animal models, which hinders their direct translation to human studies. Recently, models aiming at describing the specific cellular dynamics of human cardiac myocytes have been introduced, although limited by the scarce availability of data for model validation and calibration [70]. They aim at reproducing the electrical activity of the cell with high fidelity, at the expense of increased number of parameters: this makes them typically more computationally intensive than phenomenological models. Parameter estimation is also more difficult, as some quantities needed for model calibration may not be directly observable in non-invasive experiments.

An example: the Mitchell–Schaeffer model

The model proposed by Mitchell and Schaeffer (M-S) [71] allows capturing normal electrophysiology as well as tissue-level pathologies like cardiac dyssynchrony and minor to mild arrhythmias

$$\frac{dv}{dt} = J_{in}(v, h) + J_{out}(v) + J_{stim}(t). \tag{1.1}$$

There are two unknowns in the model: the trans-membrane potential, or voltage $v(t)$, and a gating variable $h(t)$, which models the state of the ion channels (opened / closed) that control the inward and outward currents across the cell membrane. The M-S model can therefore be seen as a "lumped" simplification of more complex, ionic models. According to Eq. (1.1), the temporal change of trans-membrane potential equals to the combined effect of the inward current, outward current, and stimulation current. J_{in} is the inward current and is computed using

$$J_{in}(v,h) = \frac{hC(v)}{\tau_{in}}, \tag{1.2}$$

where the cubic function $C(v) = v^2(1 - v)$ describes the voltage dependence of the inward current, which mimics the behavior of the fast acting ionic channel gates. τ_{in} is a time constant representing the time required for the ion channels to open. J_{out} is the outwards current, and is computed using

$$J_{out}(v) = -\frac{v}{\tau_{out}}, \tag{1.3}$$

where τ_{out} is a time constant representing the time required for the ion channels to close. Being an outwards current, it has a negative sign. Finally, J_{stim} is the stimulation current, which is applied externally by a pacemaker for instance.

The gating variable $h(t)$ is a non-dimensional variable which varies between 0 (gate is closed) and 1 (gate is open). The evolution of $h(t)$ is governed by:

$$\frac{dh}{dt} = \frac{1-h}{\tau_{open}} H(v - v_{gate}) - \frac{h}{\tau_{close}} H(v_{gate} - v), \tag{1.4}$$

where H is the Heaviside function, τ_{open} and τ_{close} are the time constants governing the duration of the phases in which the gates are open and closed, respectively, and v_{gate} is the change-over voltage.

Fig. 1.6 illustrates how these parameters relate to the shape of the action potential. Other important quantities are the action potential duration (APD) and the diastolic interval (DI), which represent the total time in which the cell is depolarized, and the time interval from cell repolarization to the next depolarization. The M-S model can produce different action potentials based on when the stimulus is provided. If the system is not starting from equilibrium because of a short diastolic interval, a shorter action potential will be generated. The relationship between APD and DI is called the restitution curve and plays a crucial role in complex arrhythmias like ventricular tachycardia or premature ventricular contractions.

The M-S model is characterized by several useful properties that make it particularly suitable for patient-specific simulations:

- The state variables are directly related to the ionic currents that cross the cellular membrane.
- The parameters are closely related to the shape of the action potential.

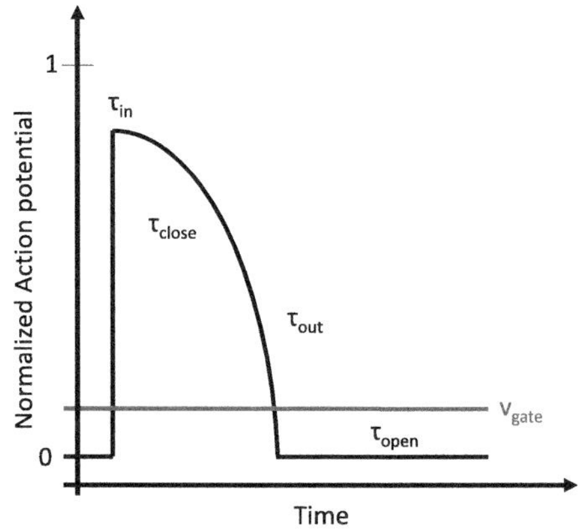

Figure 1.6. Relationships between model parameters and shape of the action potential. Model parameters can be directly related to clinical parameters.

- A closed-form relationship between model parameters and restitution curve is available, which can be used for model personalization.

Its ability to reproduce the shape of the action potential and the restitution properties of the action potential duration, combined with its computational simplicity contributed to its diffusion in the community for ventricular electrophysiology studies and in particular for parameter identification [72–74].

1.2.2 Tissue electrophysiology

The propagation of an electrical signal in the cardiac tissue can be formulated as a reaction–diffusion differential problem. In this formulation, the evolution of the electrical potential in space and time is determined by the interaction between a source term representing the generation of the action potential by the cellular component; and a diffusion term describing how the tissue properties (such as anisotropy) affect the propagation speed in different directions. Two variants are commonly considered, the monodomain and the bidomain models. *Bidomain* models explicitly account for the presence of an intra-cellular and an extra-cellular environment, each with a different electrical potential [75,76]. These models can capture specificities of each cellular domain, at the price of higher computational burden. *Monodomain* models simplify the formulation by considering only one cellular environment and computing the transmembrane potential directly.

An alternative formulation of the problem focuses on predicting the arrival time of the electrical wave at each point in the myocardium. By neglecting the explicit modeling of signal generation and propagation, this approach significantly simplifies the mathematical formulation as a traveling wave problem, expressed using an anisotropic Eikonal equation [77–79]. Such models are governed by very few parameters and they can be solved very efficiently using anisotropic fast marching methods [80,81] or graph-based techniques [82]. They are therefore suited for real-time simulations as one full-heart depolarization can be simulated in seconds. Relatively recently, advanced implementation techniques enabled to simulate wave reentry [83], which was believed to be out of reach for such models. However, because of the extreme simplifications introduced in the model, pathological phenomena like arrhythmias, fibrillations or tachycardia cannot be simulated.

One example: a monodomain model

The general monodomain formulation can be written as a parabolic partial differential equation for the trans-membrane action potential v:

$$\chi\left(C_m\frac{\partial v}{\partial t}+J_{ion}\right)=\nabla\cdot \mathrm{R}\nabla v+J_{stim}, \tag{1.5}$$

$$J_{ion}=f(h,t,v), \tag{1.6}$$

$$\frac{dh}{dt}=g(h,t,v), \tag{1.7}$$

with boundary condition $\nabla v\cdot\mathbf{n}=0$ (corresponding to the case of an isolated heart) and proper initial condition for v. χ is the cellular membrane surface-to-volume ratio, C_m is the membrane capacitance, J_{ion} are the total ionic currents from the cellular model, R is the anisotropic tissue conductivity tensor and J_{stim} is any applied stimulus current. The tissue conductivity tensor depends on the local fiber direction, $\mathbf{f}(\mathbf{x})$ to model faster conduction along the fiber direction:

$$\mathrm{R}=\sigma\left[(1-\rho)\mathbf{f}\mathbf{f}^T+\rho\mathbf{I}\right], \tag{1.8}$$

where σ is the tissue conductivity and $\rho\in[0,1]$ is the anisotropy ratio between transverse (slower) and longitudinal (faster) conductivities. The ionic currents are obtained using a set of internal gating variables y, whose changes in time are modeled using ordinary differential equations (Eq. (1.6)). When considering the M-S cellular model described above, Eqs. (1.6) and (1.7) become Eqs. (1.9) and (1.4), respectively:

$$J_{ion}=-(J_{in}(v,h)+J_{out}(v))=-\frac{hC(v)}{\tau_{in}}+\frac{v}{\tau_{out}}. \tag{1.9}$$

Modeling the electrical conduction system

An accurate EP model needs to include the description of the effect of high-speed conducting tissues in the heart [60], which determine the pattern of ventricular excitation and contraction both in normal sequences and cardiac arrhythmias [84]. Computational models have been developed to try to reproduce the tree-like structure of the Purkinje network, and to capture peculiar features of the high-speed conducting system such as retrograde propagation. This may play a role in complex scenarios such as cardiac resynchronization therapy, in which pacing electrodes may be placed in areas of the myocardium that are reached by the distal termination of the Purkinje networks. Current imaging techniques do not allow the patient-specific measurement of the heart conduction system *in vivo*, therefore this kind of models cannot be directly validated and personalized. They are typically based on anatomical information from histological studies and tuned to correctly reproduce sites of earliest activation and normal activation sequence from electrical mapping studies. An extensive literature review of this modeling approach is provided by ten Tusscher and Panfilov [84], who also propose a model in which the location of the Purkinje fiber-ventricular muscle junction (PVJ) is derived from electrical activation maps. More recent examples of anatomically realistic models of the Purkinje network include [85, 86]. An alternative, phenomenological approach can be followed, to reproduce the depolarization pattern without describing the detailed anatomy of the high-speed conducting tissue. Some studies focused on the estimation of a time delay function to apply to regions in the endocardium [87,88], or proposed an increase of the conduction velocity in the sub-endocardial tissue to represent the influence of the Purkinje system [89]. Despite its inability to describe secondary and retrograde wavefronts, this approach is adequate under the assumption that the distribution of PVJ is dense in the endocardium, so that a single wavefront is generated.

1.2.3 Body surface potential modeling

The electrical activity of the heart induces spatial and temporal changes of the electrical potential in the whole body. As the heart tissue depolarizes, a wave of positive electrical signal diffuses through the surrounding tissue and reaches the outer body surface. Electrocardiography aims at non-invasively recording such activity by measuring signals through electrodes placed on the skin. Each electrode reports the instantaneous difference between the local electrical potential and that of a reference electrode. By defining multiple leads, connections between a refer-

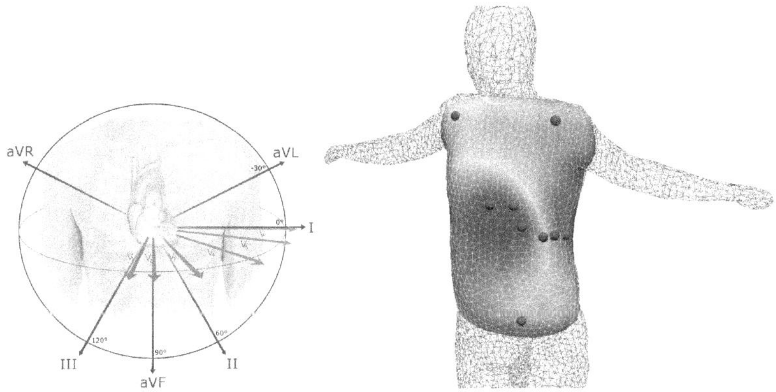

Figure 1.7. *Left:* Schematic representation of the electrocardiography leads used in 12-lead ECG *(source: Wikipedia). Right:* Idealized model of a portion of the human torso, color coded by the surface electrical signal and with overlaid location of ECG electrodes.

ence electrode and recording electrodes positioned on the body surface, it is possible to measure quantitative information about the pattern of depolarization in the body, from which cardiac activity can be inferred (Fig. 1.7). Intuitively, a positive signal on one lead means that the recording electrode is touching a region currently affected by the propagating wave. A negative signal means that the wave has passed.

In clinical practice, it is common to collect a 12-lead electrocardiogram (ECG) (Fig. 1.7), which consists of [90,91]

- three limb leads (I, II, III): bipolar leads, obtained by mutual connection of three electrodes placed on each arm and on the left leg;
- three augmented limb leads (aV_R, aV_L, aV_F): unipolar leads obtained from each limb lead considering as reference the mean of the remaining limb potentials
- six precordial leads (V_1 to V_6): unipolar leads obtained from electrodes in precordial position, using as reference the mean of the three limb potentials.

This measurement strategy allows an easy and reproducible setup, and can provide insights in the location of electrical events in the heart, although with some degree of uncertainty due to the relatively sparse sampling of the body surface potentials [92].

More detailed characterization of the electrical signal on the body surface can be obtained by multielectrode body surface potential mapping (BSPM) [93,94]. The additional information provided by adding a significant number of measurement points has been shown to enable the identification of regional cardiac events, although with limited spatial accuracy due to the intrinsic smoothing and volume averaging effect of surface measurements (as compared for instance to invasive surface mapping of the heart). Furthermore, noninvasive electrocardiographic imaging aims at using BSPM to compute epicardial potentials by solving the inverse elegrophysiology problem [95–97]. By focusing on the reconstruction of epicardial potentials, this approach has the advantage of potentially leveraging the proven effectiveness of epicardial surface mapping techniques for the localization of cardiac electrical events; however, accurate reconstruction can be challenging in real clinical scenarios [98].

Bidomain modeling of the coupled heart-body system

Body surface potentials (BSP) can be directly modeled by extending the tissue-level electrophysiology model with a component representing the body as an electrically conductive medium. In this case, the heart is electrically coupled with the surrounding tissue in the torso. Therefore, the modeling hypotheses leading to the definition of the monodomain model are not adequate, since by ignoring the extra-cellular compartment the model cannot account for electrical coupling with other tissues. The bidomain model has thus been the modeling framework of choice for BSP simulations. By recalling the definition of the transmembrane potential as $v = \phi_i - \phi_e$, with ϕ_i and ϕ_e representing the intra- and extracellular potential respectively, the bidomain model reads:

$$\begin{cases} \chi\left(C_m \frac{\partial v}{\partial t} + J_{ion}\right) = \nabla \cdot \mathrm{R}_i \nabla v + J_{stim}, \\ \chi\left(C_m \frac{\partial v}{\partial t} + J_{ion}\right) = -\nabla \cdot \mathrm{R}_e \nabla v + J_{stim}, \end{cases} \tag{1.10}$$

with the same meaning of the symbols in Eqs. (1.5)–(1.7) and where R_i and R_e represent the intra- and extracellular electrical conductivity tensors, respectively. In this case, the boundary conditions express the absence of current flow from the intracellular compartment to the surrounding tissue (the torso): $\mathrm{R}_i \nabla \phi_i \cdot \mathbf{n} = 0$, where $\mathbf{n}$ is the epicardial surface normal. Current balance is guaranteed at the interface between the extracellular compartment and the torso:

$$(\mathrm{R}_e \nabla \phi_e) \cdot \mathbf{n} = -(\mathrm{R}_o \nabla \phi_o) \cdot \mathbf{n}, \tag{1.11}$$

where R_o and ϕ_o are the conductivity tensor and electrical potential in the torso. Finally, the torso can be modeled as a passive electrical medium, for which the spatial distribution of the potential is described by Laplace's equation:

$$\nabla \cdot \mathrm{R}_o \nabla \phi_o = 0. \tag{1.12}$$

This problem formulation allows to define a direct relationship between the parameters of the electrophysiology models at the cellular and tissue level and signals that can be acquired noninvasively on the surface of the torso. Based on this, parameter identification strategies can be designed to recover the optimal parameterization of the electrophysiology models so that patient-specific ECG readings can be reproduced. Examples and more extensive discussion of this application are provided in section 5.1.

Without changing the coupling framework here presented, the different components can be modified and adapted to the requirement of the specific application. For instance, the torso can be described as a non homogeneous medium, accounting for the presence of multiple tissues with different conductive properties. This can potentially enable higher fidelity in the modeled ECG traces [92]. The tissue electrophysiology model can be simplified to an extended monodomain model [99] or even to an Eikonal model, with proper assumptions on the relationship between the transmural and extracellular electrical potentials. One example of coupled heart-torso model based on the monodomain model is presented in section 2.2.4.

1.3 Biomechanics modeling

The goal of a biomechanical model of the heart is to capture the active contraction and passive relaxation of the muscle, triggered by the cardiac electrophysiology model as described in the previous section and under the constraints determined by the surrounding tissues. The myocardium is an active, non-linear, orthotropic, visco-elastic tissue, bi-directionally coupled with cardiac electrophysiology [5,47,100]. Its constitutive law, which describes its macroscopic behavior, comprises an active component, namely the relationship between the stress generated by the muscle and the resulting contraction, and a passive component, the underlying elasticity of the tissue due to the myocytes and extracellular matrix. In practice, the active contraction is viewed as a transient external force that makes the myocardium contract, while the passive properties of the tissue are modeled such that they ensure realistic motions due to the internal forces [100]. The Hill–Maxwell framework [101,102] is traditionally used to integrate the action of these forces. Fig. 1.8 shows one Hill–Maxwell model of the myocardium [103] with three elements. The resistance W_e represents the passive work. τ_c is the macroscopic active stress, controlled by the electrical signal u, while E_s is an additional elastic element that allows modeling of the isometric behavior (when active stress results in no strain). Dissipation due to friction and viscous effects can be added into both the active and passive models (μ and respectively η in the figure).

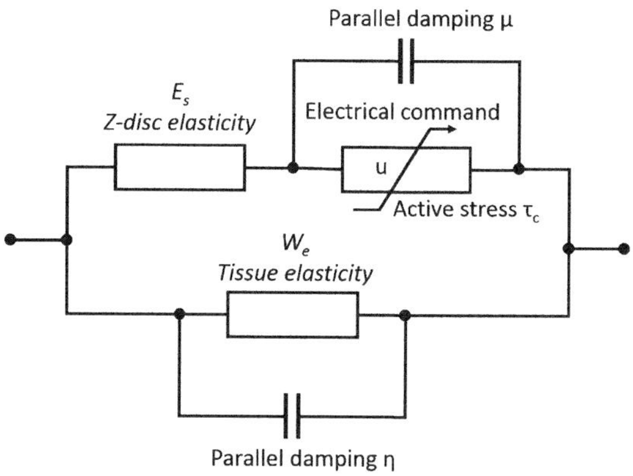

Figure 1.8. Illustration of an advanced version of the Hill–Maxwell rheological model of cardiac biomechanics.

To compute cardiac biomechanics, one solves Newton's equation of motion:

$$\mathrm{M}\ddot{\mathbf{u}} + \mathrm{D}\dot{\mathbf{u}} + \mathrm{K}\mathbf{u} = \mathbf{f}_a + \mathbf{f}_p + \mathbf{f}_b, \quad (1.13)$$

$\mathbf{u}$ denotes the displacement vector of a material point at time t, $\dot{\mathbf{u}}$ is the velocity and $\ddot{\mathbf{u}}$ the material point acceleration. M is the mass matrix, D is the damping of the system, often chosen to be of Rayleigh type, and K is the anisotropic stiffness matrix. $\mathbf{f}_{bp}$ captures the ventricular blood pressure, applied as a boundary condition to the endocardial surface and computed by both myocardium biomechanics and hemodynamics models. $\mathbf{f}_b$ accounts for external boundary conditions and $\mathbf{f}_a$ is the active force generated by the depolarized cells. Eq. (1.13) is traditionally solved using spatio-temporal discretization schemes like the finite element method (FEM) [104] with (semi-)implicit [105] or explicit [106] time integration.

The following sections provide a brief survey of possible modeling choices for the passive and active stress components, as well as the integration of boundary conditions and constraints. In [47], the authors provide a review focused on how computational models of the heart can be integrated with medical imaging data. Niederer et al. [107] present a modern look on computational modeling, discussing salient points like data assimilation and model personalization in presence of various pathologies. Finally, although not described in this book, another area of great research interest is the study of long-term myocardium remodeling due to growth or pathologies. We refer the readers to [108–111], and references therein, as notable examples.

1.3.1 The passive myocardium

Experimental testing showed that the passive myocardium is an essentially incompressible orthotropic material, characterized by distinct material responses in three mutually orthogonal planes defined by the fibers and fiber sheets. A large variety of models have been proposed to simulate these properties [43,112,113]. Constitutive laws are often formulated as an energy-strain functional in polynomial or exponential form [102,113]. In [114], the authors proposed an early model of the myocardium using a transverse isotropic strain energy density function. This model, referred to as the Guccione law, is still used in benchmarks [115]. More complex models have been proposed to include the effects of myocardial sheets, assumed to be involved in myocardium thickening during systole. A first category of models are those based on the pole-zero technique, originally proposed by Hunter and colleagues [116] and then extended by Niederer et al. [117]. In this approach, the microstructure of the myocardium is modeled with three independent axes (fibers, sheets and normals axes). A pole-zero formulation is then used to model each axis with an independent, separate pole, thus accounting for the different strain behavior in each axis with a formulation shown to be more numerically stable than traditional alternatives. In [118], the authors analyzed closely the fiber/fiber sheet mechanism and proposed an exponent law, now referred to as the Costa law and largely adopted in the community. In [119], the authors presented a quantitative comparison of the most common models at the time, in terms of prediction accuracy with respect to ex-vivo experiments. In these experiments, the Costa law tended to outperform the other models. More recently, the structurally-based Holzapfel–Ogden (HO) model [120] has gained popularity among the community, arguably making it the current state-of-the-art. Contrary to the phenomenological constitutive laws of Costa and Guccione, the HO model is derived from considerations on the microstructure of the tissue, and not by fitting exponential functions to stress-strain relationship observed experimentally. It is also easier to implement and the parameters are directly related to the structural function of the muscle. Other groups rely on more standard models like Mooney–Rivlin [103,121] or corotational linear elasticity [6], both simpler and more computationally efficient. However, advanced numerical schemes now allow fast computation of hyper-elastic models, making the less accurate linear models obsolete.

The strain energy function of the commonly used Guccione and Costa laws is defined by an exponential relation of the form $\psi = c(\exp(dQ) - 1)$, where c and d are stiffness parameters, and Q is a functional expression of the strain that models tissue nonlinearity. The model proposed by Guccione and colleagues considers the anisotropy along the fiber orientation:

$$Q = b_1 E_{ff}^2 + b_2(E_{cc}^2 + E_{rr}^2 + 2E_{cr}E_{rc}) + 2b_3(E_{fc}E_{cf} + E_{fr}E_{rf}).$$

In this equation, E_{ij} are the components of the Lagrangian strain tensor $\mathrm{E} = (E_{ij})$. The indices f, c and r denote the fiber direction **f**, the cross-fiber direction **c** and the radial direction **r** respectively. This model depends on three parameters, b_1, b_2 and b_3. Costa and colleagues improved that model to include the myocardium sheets:

$$Q = c_1 E_{ff}^2 + c_2 E_{ss}^2 + c_3 E_{nn}^2 + 2c_4 E_{fs}E_{sf} + 2c_5 E_{fn}E_{nf} + 2c_6 E_{sn}E_{ns},$$

where the indices f, s and n denote the fiber, sheet and normal directions. In practice, the parameters b_i and c_i are not adjusted to patient data but rather fixed from ex-vivo experiments. The modeler only adjusts the parameters c and d of the energy function to fit the model to the measurements.

An overview of the Holzapfel–Ogden model

Let us first introduce a few elements of continuum kinematics. Given a differentiable material deformation function ϕ, which maps the 3D domain of interest (e.g. myocardium) from time $t = 0$ to any other time instant t, the basic deformation variable for the description of the local kinematics is the deformation gradient $\mathrm{F} = \nabla\phi$. This has a positive Jacobian, $J = \det \mathrm{F}$, with $J = 1$ for incompressible tissue. F allows to compute the right (C) and left (B) Cauchy–Green tensors, defined by

$$\mathrm{C} = \mathrm{F}^T\mathrm{F}, \tag{1.14}$$

$$\mathrm{B} = \mathrm{F}\mathrm{F}^T. \tag{1.15}$$

The Green–Lagrange strain tensor then writes

$$\mathrm{E} = \frac{1}{2}(\mathrm{C} - \mathrm{I}), \tag{1.16}$$

where I is the identity tensor. The principal (isotropic) invariants of C (and also of B) are defined by $I_1 = tr(\mathrm{C})$, $I_2 = \frac{1}{2}[I_1^2 - tr(\mathrm{C}^2)]$ and $I_3 = det(\mathrm{C})$.

The HO model takes into account the observation that the cardiac tissue shows different behaviors whether it is stretched along the fiber, along the sheets or in the fiber-sheet plane normal. For all these conditions, the stress-strain relationship is exponential. Based on these considerations, the HO energy function writes:

$$\begin{aligned}\psi =&\frac{a}{2b}exp[b(I_1-3)]\\&+\sum_{i=\{f,s\}}\frac{a_i}{2b_i}\left\{exp[b_i(I_{4i}-1)^2]-1\right\}\\&+\frac{a_{fs}}{2b_{fs}}\left[exp(b_{fs}I_{8fs}^2)-1\right].\end{aligned} \tag{1.17}$$

Here a, b, a_i, b_i, a_{fs}, b_{fs} are tissue parameters with $i=\{f,s\}$, the a parameters having dimension of stress, whereas the b parameters are dimensionless. Furthermore, the following deformation invariants are defined: $I_{4i}=\mathbf{i}^T\mathbf{C}\mathbf{i}$, with $\mathbf{i}=\{\mathbf{f},\mathbf{s}\}$, and $I_{8fs}=\mathbf{f}^T\mathbf{C}\mathbf{s}$. To model near incompressible materials like the heart it is customary to add the term $d_1(J-1)^2$, with d_1 homologous to a bulk modulus. Note that the second and third terms of the energy function (Eq. (1.17)) are null when the tissue is under compression, i.e. $I_{4i}<1$, since fibers do not show resistance to compression. In summary, the HO model is governed by nine parameters (Table 1.1), obtained from the experiments reported in [113].

A simplified version of the HO model is often chosen, where the contributions of the fiber sheets are neglected. Indeed, current imaging limitations do not allow in-vivo assessment of myocardial fibers, and even less fiber sheets. Furthermore, the impact of the sheets on the overall cardiac function is still not well understood. Therefore, the transverse isotropic version of the Holzapfel–Ogden model is often preferred for patient-specific simulations, having less parameters to identify. The modified strain energy function becomes:

$$\psi=\frac{a}{2b}exp[b(I_1-3)]+\frac{a_f}{2b_f}\left\{exp[b_f(I_{4f}-1)^2]-1\right\}.$$

Table 1.1 Parameters and default values of the hyper-elastic model of cardiac myocardium proposed by Holzapfel–Ogden (values from [113]).

a (kPa)	b	a_f (kPa)	b_f	a_s (kPa)	b_s	a_{fs} (kPa)	b_{fs}	d_1
0.496	7.209	15.193	20.417	3.283	11.176	0.662	9.466	496

1.3.2 The active myocardium

The active contraction of the myocardium is initiated by cardiac electrophysiology. As the myocytes depolarize, calcium (Ca^{2+}) channels open and Ca^{2+} ions cross the cellular membrane to bind to the sarcoplasmic reticulum (SR), activating the so-called calcium-induced–calcium-release (CICR) mechanism. According to the sliding filament hypothesis [42], the surplus of Ca^{2+} released inside the cell by the SR activates the sarcomeres, in particular the actin-myosin myofilaments (Fig. 1.9 and Fig. 1.10). When the cell repolarizes, the calcium concentration within the myocyte comes back to its initial values, and the sarcomeres relax [42]. An active contraction model should therefore be connected to the cardiac electrophysiology model and capture both contraction and relaxation phases. As for cardiac electrophysiology, multiple models have been proposed, with various levels of details. Three categories can be distinguished: ionic, phenomenological and lumped models.

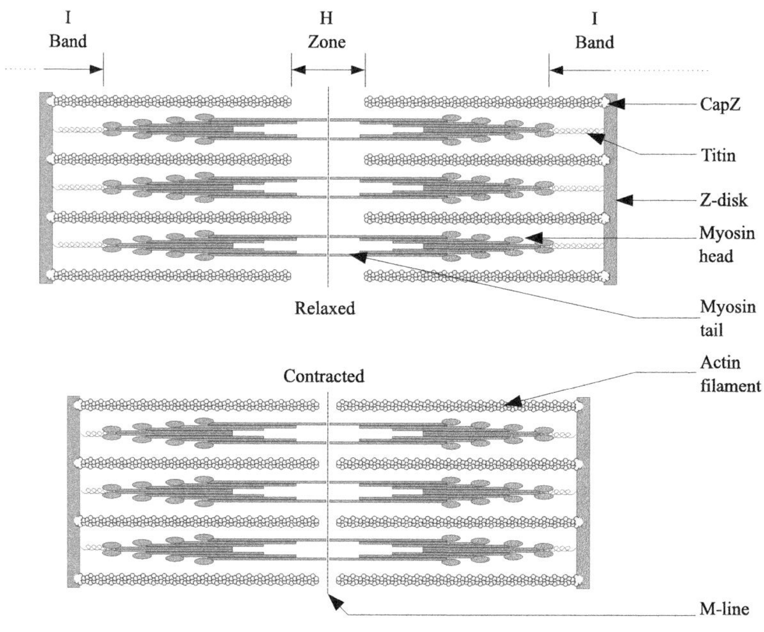

Figure 1.9. Diagram illustrating the structure of a sarcomere, with the different myofilaments. (*Source: Wikipedia.*)

Ionic models

Biophysical models simulate the ion interactions and the actin-myosin bindings that generate the cardiac motion [14,100,

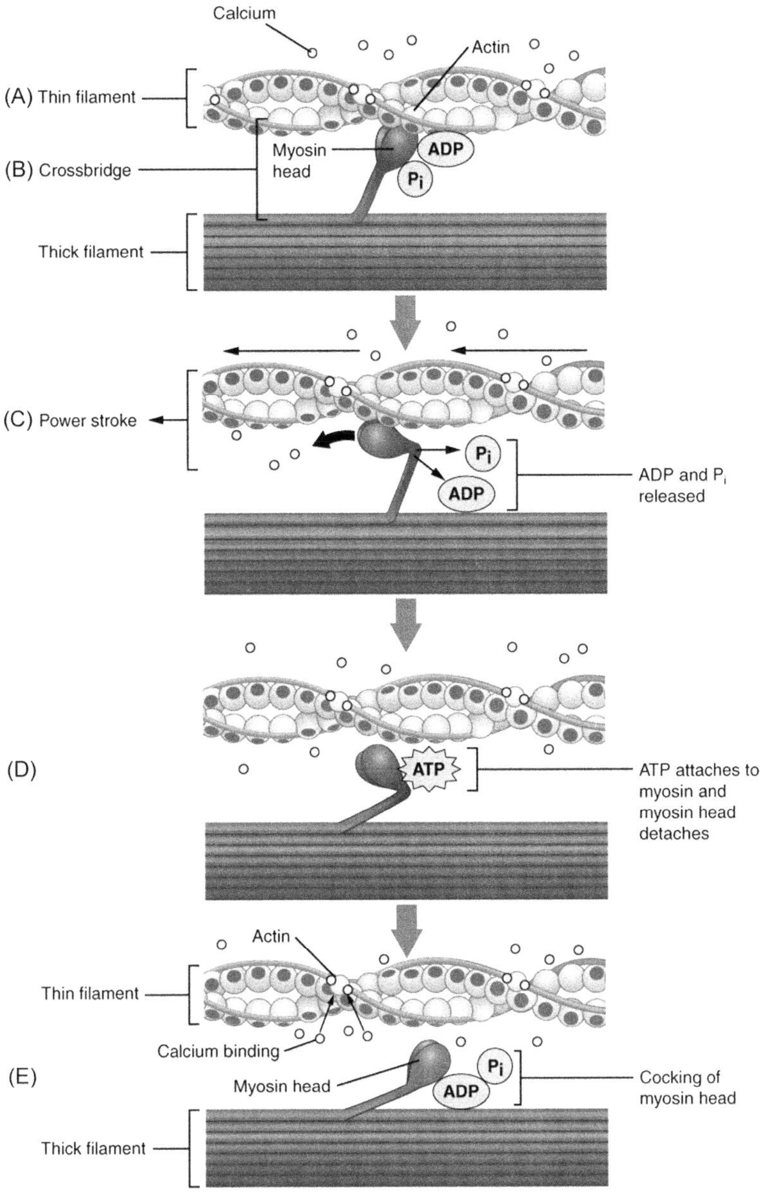

Figure 1.10. Diagram illustrating the different steps of the sliding myofilaments mechanism. (*Source: Wikipedia.*)

117,122]. They were developed from experimental studies on ex-vivo animal hearts, often at the cellular level, to discover the mechanisms of myocytes physiology at the molecular and cellular scale.

The Rice model [123] is commonly used in recent simulation studies. It has been further augmented with metabolic and energetic considerations [124]. This type of models captures most of the known cellular and molecular mechanisms involved in myofilament function, from the CICR mechanisms to troponin function and ion binding to force generation. These models can comprise more than 40 ordinary differential equations that need to be integrated. Scaling up these models at the organ level is therefore challenging and computationally demanding (40+ equations per mesh node), although not impossible [123,125], provided they are coupled with an electrophysiology model that can calculate directly the intra-cellular Ca^{2+} concentration. Nonetheless, the very large number of free parameters (40+) and their direct link to ionic and molecular properties make these models difficult, if not impossible, to personalize from standard of care clinical data.

Phenomenological models

The idea behind phenomenological models is to provide an integrated description of the biological mechanisms from the myofilaments to the organ [103,126]. The transition from one spatio-temporal scale to another is achieved mathematically, using mean field theory for instance [103], ultimately resulting in a set of simplified equations that are controlled by fewer parameters (usually four to five parameters). Sermesant et al. [45] proposed a simplified version of these models, with analytical integration for model-based image analysis. A multi-scale model that also considers energy exchange during the heart beat was then proposed in [103], to ensure the balance between oxygen supply and energy consumption. Details of a phenomenological model used for patient-specific simulations are given in section 2.3.2.

Lumped models

Lumped models are analytical models of the fiber contraction that do not consider spatial variability – therefore not requiring 3D meshes to be solved. They focus on an averaged myocyte response, with a small number of bulk response laws characterizing the contraction [127]. These models can be solved very efficiently but they cannot capture regional abnormalities of the myocardium in patients, like scars or localized fibrosis for instance. As the research community focuses more on patient-specific simulations, lumped models regained interest for their computational efficiency. In particular, hyper-reduced models have been developed as surrogates of more complex 3D models, in order to achieve fast model personalization [128,129].

All the models mentioned above provide the active stress τ_c generated by the sarcomeres (Fig. 1.8). Following the spatial discretization scheme, the stress is integrated to obtain the nodal forces $\mathbf{f}_a$ that appear on the right-hand side of Newton's law of motion (Eq. (1.13)). Section 2.3 presents a concrete example of active stress and its numerical derivation in an FEM solver.

1.3.3 The virtual heart in its environment: boundary conditions

Heart function depends on the constant interactions between the myocardium, neighboring organs, intra-cardiac blood flow and global circulation. These interactions are modeled as boundary conditions to the biomechanical model described in the previous sections. As formalized in Eq. (1.13), two types of boundary conditions need to be defined: the pressure force exercised by the blood flow on the endocardial surfaces, denoted $\mathbf{f}_{bp}$, and the attachments of the myocardium to the other organs, denoted $\mathbf{f}_b$.

Endocardial pressure

Intra-cardiac blood flow imposes stresses on the endocardium, which can vary significantly throughout the four phases of the cardiac cycle: isovolumic contraction, ejection, isovolumic relaxation, filling. Researchers investigated fluid-structure interaction (FSI) methods to understand and quantify the 3D patterns of the blood flow inside the chambers, along with the spatially varying stresses on the endocardial surfaces [130,131]. To that end, a complete anatomical model of the heart, with all chambers and valves, must be available. These modeling approaches can be very detailed and, correspondingly, complex to solve, with coupled systems controlled by large sets of parameters.

In order to achieve faster computations, researchers proposed simplified FSI models in which the pressure is assumed uniform in each chamber (i.e. it does not vary spatially). The computed pressure is applied to the surface of the biomechanical model as a normal force ($\mathbf{f}_{bp}$ in Eq. (1.13)). The time-varying pressure depends then on 1) the force generated by the myocardium and on 2) the circulatory system. The latter is modeled using lumped parameters models (LPM) like those presented in [78,132,133] (see next section), or directly prescribed by the user [45]. Two approaches are possible to achieve the biomechanical/hemodynamics coupling:

- **Pressure-driven models**: the pressure is an unknown of the coupled system, and hence needs to be solved [121]

- **Flow-driven models**: the pressure is prescribed to the biomechanical solver and is known at any time [45].

Pressure-driven models are more complex to solve but offer a larger flexibility and fidelity, as they can naturally model valve insufficiency, stenosis, and other hemodynamics diseases. Section 2.3 presents an implementation of such a model. In brief the ventricular hemodynamics is modeled using a homogeneous pressure field [6]. The arterial pressure is represented using a lumped Windkessel model, while the atrial pressure is represented using an active elastance model [133]. Finally, the valves, which control the blood flow through the chambers and hence the transition from one cardiac phase to the other, are represented as 0D dynamical systems, functionally coupled with the ventricles, arteries and atria through the pressure variable.

Attachment to neighboring vessels and tissue

The second type of boundary conditions model the interaction of neighboring organs with the myocardium. These boundary conditions have been found crucial for realistic simulations and to avoid non-physiological apex motion or rocking of the ventricles [134]. First, heart ventricles and atria are connected to the vessels, which creates additional stiffness in the insertion regions. A common way to model these effects is to add stiff springs $\mathbf{f}_{vessels}$ at the vessels insertion points [6,45,134]. Second, a pericardium constraint has also been used by some groups to achieve more realistic deformation [6,135,136]. The idea consists in constraining the epicardial motion either through stiff springs or by using an explicit, contact-based, friction-less pericardial force $\mathbf{f}_{peri}$, whose domain is derived from the epicardial surface at mid-diastole [135]. The total boundary condition $\mathbf{f}_b = \mathbf{f}_{vessels} + \mathbf{f}_{peri}$ is then injected into Eq. (1.13). Implementation details are provided in Section 2.3.

1.4 Hemodynamics modeling

The mechanical coupling between the heart and the blood circulatory system is a crucial aspect of heart function. Venous return to the atria (through the inferior and superior vena cava on the right side and through the pulmonary veins on the left side) provides the preload conditions to the cardiac pump and determines the stroke volume of the ventricles. Mean arterial pressure (in the aorta on the left side and in the pulmonary artery on the right side) represents the afterload conditions and determines the amount of work that the heart has to exert in one heart beat. The dynamics of blood in the heart chambers and coronary vessels characterizes physiological or pathological conditions such as valvular defects, myocardial infarction, thrombosis, cardiomyopathies.

In normal physiological conditions, an increased demand of blood flow from the peripheral organs causes vasodilation and a corresponding reduction in peripheral resistances. This results in increased flow rate of circulating blood, even without changes to the mean arterial pressure, and therefore an increase in the amount of blood filling the atria and ventricles at the end of each cardiac cycle. According to the Frank–Starling's law, the heart can match increased end diastolic volume of the heart chambers with an increased stroke volume, thus adapting to the blood flow demand from the peripheral organs [137]. The underlying mechanism responsible for such adaptation is the length-tension relationship of the myocardium, whereas the myofibers can produce greater tension if their initial length is stretched beyond their normal resting length [101].

Increases in the mean arterial pressure can be caused by a range of factors such as smoking, lack of physical activity, obesity, diet and others [138]. This translates in increased ventricular pressure during the systolic phase, which requires the ventricles to express an increased level of tensile stress. As a result, the work required from the ventricle to guarantee the cardiac output increases. This also explains why chronic hypertension may promote cardiac hypertrophy: a thicker myocardium can better account for the increased work demand [139].

Modeling hemodynamics in the heart chambers and in the circulatory system allows to better understand the patient-specific context in which the heart is operating. Depending on the focus of interest, different modeling approaches can be considered [140]. Detailed analysis of the 3D fluid dynamics of blood can be an important tool in studying clinically relevant problems such as valve mechanics (in particular for diseased valves) [141–143], coupling of the myocardium with devices such as ventricular assist devices [144–146], analyzing the effect of cardiac interventions [147–149].

Reduced order models may be a more practical and convenient choice for the monitoring of spatial-averaged quantities such as blood pressure and flow rate, for instance for the computation of pressure-volume loops in which the model provides an estimation of intracardiac pressure removing the need for invasively measured pressures [150,151]. They are also used in combination with detailed hemodynamics models to describe larger portions of the cardiovascular system, while keeping the computational complexity under control. This modeling approach is commonly referred to as *geometrical multiscale modeling* and aims at focusing the use of detailed and computationally complex models for the representation of spatially limited regions of interest while using reduced order models to account for the mechanical coupling with the remainder of the circulation. An extensive review of this modeling approach is presented in [152].

In the following we briefly discuss different approaches for modeling hemodynamics, providing a few examples.

1.4.1 Reduced order hemodynamics

To model the relationship between local hemodynamics (in the heart) and global hemodynamics (in the rest of the circulatory system) one would ideally use a Whole Body Circulation (WBC) model which includes, besides the lumped cardiac system, integrated systemic and pulmonary circulations. This can be done using a reduced order flow and pressure model [150,151,153–160], such as the one illustrated in Fig. 1.11 below. In this model, the variables of interest are blood flow rate Q and blood pressure P, and different parts of the circulation are described as different independent components, connected to each other according to basic conservation principles (flow rate and pressures are conserved at the interface between components). This is formally analogous to the modeling approach used for electrical network systems, when we consider electrical current in place of blood flow rate and electrical potential in place of blood pressure [153]. By applying Kirchhoff's circuit laws, a differential algebraic system of equations is obtained describing the temporal dynamics of the variables of interest at each interface point. A more detailed description of each component used in the circulation model follows.

Ventricular model

The simplest surrogate for the three-dimensional biomechanical model of the heart would be a lumped (0D) version, able to describe the main aspects of heart function while allowing fast personalization of the model parameters. One example of such model features a time-varying elastance function governing the relationship between ventricular pressure and volume [161–163]:

$$P(t) = E(t)(V(t) - V_0)(1 - K_s Q), \tag{1.18}$$

where E is the time-varying elastance, V is the ventricular volume, V_0 is the dead volume of the ventricle, $Q = \frac{dV}{dt}$ is the total ventricular flow, and K_s a constant. We have used here the simplified model presented in [163].

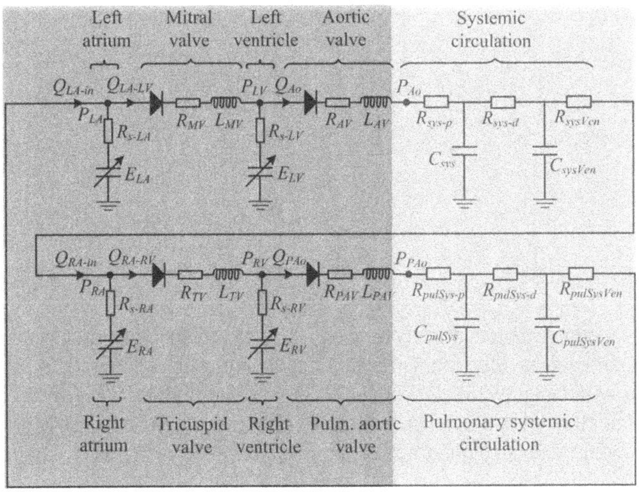

Figure 1.11. Lumped parameter model representing the whole body circulation. Heart systems are in red (dark gray in print version), systemic and pulmonary circulations in blue (light gray in print version).

Valve model

The lumped valve modules receive temporal pressure information from both sides (ventricle on one side and artery or atrium on the other) and provide corresponding flow values. They simulate the opening and closure of the valve based on the pressure gradient between the two sides of the valve. Following [151,163] one can model the pressure drop ΔP due to flow as:

$$P_{in} - P_{out} = R_{valve} Q + B_{valve} Q|Q| + L_{valve} \frac{dV}{dt}, \tag{1.19}$$

where P_{in} and P_{out} represent the pressures at the inlet and respectively the outlet of the valve. The sign convention varies for the different valves. Valves located at the ventricular inflow (mitral valve for the left ventricle, tricuspid valve for the right ventricle) see a positive flow entering the ventricle; valves located at the ventricular outflow (pulmonary valve for the right ventricle, aortic valve for the left ventricle) see a positive flow exiting the ventricle.

The pressure losses on the right-hand side of Eq. (1.19) are due to a Poiseuille (viscous) component $R_{valve} Q$, a convective inertia component dubbed Bernoulli loss $B_{valve} Q|Q|$ and a transient inertia term $L_{valve} \frac{dV}{dt}$. The coefficients depend on the geometry of the opening as follows: the viscous resistance coefficient $R_{valve} \sim 1/A^{1.5}$, the Bernoulli coefficient $B_{valve} \sim 1/A^2$ and the transient inertia coefficient $L_{valve} \sim l/A$ where l is an effective length pro-

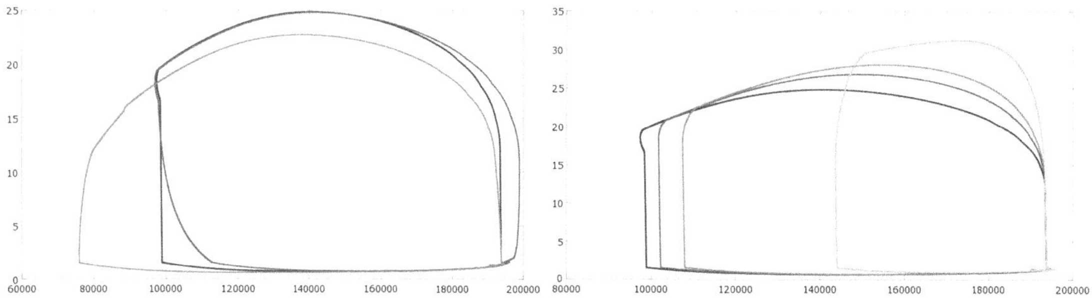

Figure 1.12. Computation examples using the lumped valve to model pathology like insufficient and stenotic valves. *Left panel*: LV PV loops in the case of regurgitant valves. Blue (dark gray in print version) – no regurgitations, red (light gray in print version) – mitral regurgitation, green (mid gray in print version) – aortic regurgitation. *Right panel*: LV PV loops for aortic stenosis of increasing degrees. Blue (dark gray in print version) – normal, green (mid gray in print version) – mild, red (gray in print version)– moderate, cyan (light gray in print version) – severe. The abscissa units are mm^3 and the ordinate units are kPa.

portional to the valve height. Here A is the effective opening area of the valve, whose variation over time is modeled as:

$$A(t) = A^{max}[(M^{sten} - M^{reg})\phi(t) + M^{reg}],$$

with A^{max} being the maximal opening area, and with the phase of the valve $\phi(t)$ being defined as a differentiable function varying between 0 (closed valve) and 1 (open valve). Its dynamics is controlled by the pressure gradient as follows:

$$\frac{d\phi}{dt} = \begin{cases} (1-\phi)K^{open}\Delta P, & \text{if } \Delta P > 0 \\ \phi K^{close}\Delta P, & \text{if } \Delta P < 0 \end{cases}$$

where K^{open} and K^{close} are opening and closing rate coefficients, which can be personalized based on extracted valve kinematics, if available. We note that the model valve opens and closes at a faster rate when ΔP is greater, whereas for a fixed ΔP, valve motion slows down as it approaches the fully open or fully closed position. The constants M^{sten} and M^{reg} characterize the stenotic and regurgitation properties of the valve, and lie between 0 and 1. They are 1 and respectively 0 for normal hearts, with values strictly between 0 and 1 used to model the various pathology combinations. An example of the variation in the PV-loops that can be obtained by varying these parameters is provided in Fig. 1.12.

The valve phase equations are simple ODEs that can be solved accurately with second-order accuracy methods (e.g. Euler, Runge–Kutta, etc.). Eq. (2.22), which ensures the coupling of both valve modules on the same side of the heart and the biomechanics heart solver, can similarly be solved with fully or semi implicit discretization methods. In practice, as shown by [151], one can obtain physiologically meaningful results by considering only one lumped resistance component, such that for example the pressure drop across the valve becomes $\Delta P = RQ$. In such case, Eq. (2.22) can be written

$$\frac{dV}{dt} = \frac{\Delta P^{arterial}}{R_1} + \frac{\Delta P^{atrial}}{R_2} - \mu \frac{dP}{dt} \tag{1.20}$$

and it can be used to determine the state parameter κ, which for each valve verifies $R_i = \kappa / A_i^{1.5}$, where R_1 corresponds to the arterial valve, and R_2 corresponds to the atrial valve. The algorithmic steps for valve model update then become the following (for each side of the heart):

Algorithm 1 Valve update algorithm.

1. from the pressure drop across each valve find its opening phase and its effective opening area
2. from Eq. (1.20) find the state parameter κ
3. find each valve flow rate and send it to its corresponding remote module (arterial Windkessel or atrial module)

Arterial model

The arterial pressure variation can be modeled using a 3-element Windkessel model [164], which takes as input the arterial flow and returns the pressure within the artery. The first element of the model is a peripheral resistance R_p, which accounts for the distal resistance of the circulatory system mainly due to the small vessels. The compliance C accounts for the elasticity of the arterial walls, whereas the characteristic resistance R_c accounts for the blood mass and for the compliance of the artery proximal to the valves. Let $Q_{ar}(t)$ be the arterial flow at time t, defined as positive when exiting the ventricle, $P_{ar}(t)$ be the arterial pressure at time t and P_r be a constant low pressure of reference (for example the pressure of the remote venous system for the left side circulation). If the model is a closed whole body circulation system, one needs to enforce that P_r coincides with the corresponding atrial pressure. When the blood flows into the arteries ($Q_{ar}(t) > 0$), during ejection, the 3-element model is:

$$\frac{dP_{ar}}{dt} = R_c \frac{dQ_{ar}}{dt} + \left(1 + \frac{R_c}{R_p}\right) \frac{Q_{ar}}{C} - \frac{P_{ar} - P_r}{R_p C}.$$

When the valves are closed, or the flow is regurgitant, the appropriate sign change for the flow must be enforced, along with a possible update to the resistance and compliance parameters – corresponding to a "depleting" rather than a "filling" reservoir model. These simple ODEs are easily integrated using for example first-order (Euler) methods.

Atrium model

Atrium contraction, which happens just after diastasis and before systole, optimizes ventricular filling. Because simulating atrial contraction explicitly in 3D may be computationally demanding, a common approach is to rely on lumped models that mimic the raise of ventricular pressure due to atrial contraction. While some simplified models consider atrial pressure constant, a common strategy consists in using phenomenological models of atrial pressure based on sigmoid functions, e.g. [45]. More predictive elastance models have also been proposed to capture the interactions between atrial volume, pressure, tissue stiffness and circulatory system [133].

Atrial contraction can be modeled using a lumped time-varying elastance model. For both atria, the pressure is computed according to the equation (for left atrium LA, with similar equations for the right atrium):

$$p_{LA} = E_{LA}(V_{LA} - V_{LA,rest}),$$

where the elastance E_{LA} and the rest volume V_{LA} are:

$$\begin{aligned} E_{LA} &= y_a(E_{LA,max} - E_{LA,min}) + E_{LA,min}, \\ V_{LA} &= (1 - y_a)(V_{LA,rd} - V_{LA,rs}) + V_{LA,rs}. \end{aligned}$$

In these equations, $E_{LA,max}$, $E_{LA,min}$, $V_{LA,rd}$ and $V_{LA,rs}$ are free parameters of the model (maximum and minimum elastance, diastolic and systolic volumes at zero pressure respectively). Minimum and maximum elastance parameters enable to set the peak systolic and diastolic stiffness, which then controls atrial pressure based on the current volume.

A simple model of atrial activation enables controlling atrial volume. The activation function y_a is defined by

$$y_a = \begin{cases} 0.5\,[1 - \cos(2\pi t_{atrium}/t_{twitch})], & \text{if } t_{atrium} < t_{twitch} \\ 0, & \text{if } t_{atrium} > t_{twitch} \end{cases}$$

where t_{twitch} is the duration of the atrial contraction and:

$$t_{atrium} = \begin{cases} mod(t - t_{active} + \Delta t_{PR}, t_{cycle}), & \text{if } t \geq t_{active} - \Delta t_{PR} \\ 0, & \text{if } t < t_{active} - \Delta t_{PR}. \end{cases}$$

Atrial contraction is synchronized with the ventricular electrophysiology model through a time-shift parameter Δt_{PR} corresponding to the ECG PR interval.

Finally, the volume of the left atrium is given by integrating the ordinary differential equation (ODE):

$$\frac{dV_{LA}}{dt} = Q_{pv} - Q_{mitral},$$

where Q_{mitral}, the flow through the mitral valve, is treated as an independent variable which balances by mass conservation the volume variation of the 3D ventricle and the aortic flow. Q_{pv}, the flow through the pulmonary veins, is given by

$$Q_{pv} = (p_{pv} - p_{LA})/R_{pv},$$

where p_{pv} is the pulmonary vein pressure and R_{pv} the resistance of the pulmonary veins.

Venous circulation

In the systemic and pulmonary venous circulation the impedance of the proximal part of the vessel can be neglected compared to the peripheral resistance. A simple approach is to then use a two-element Windkessel model consisting of a compliance and a resistance. For the systemic venous circulation the equations are (refer also to Fig. 1.11):

$$\frac{dP_{ven}}{dt} = \frac{Q_{ven}}{C_{sysVen}} - \frac{P_{ven} - P_r}{R_{sysVen}C_{sysVen}},$$

where C_{sysVen} is the venous systemic compliance, R_{sysVen} is the venous systemic resistance. Similar equations can be employed for the pulmonary venous circulation.

1.4.2 3D hemodynamics

The fluid dynamics in the heart is very complex. The geometry of the chambers, as well as their fast motion, induces intricate flow patterns. Blood can move through the valves with speed in excess of 1 m/s [165] in normal physiological conditions, and even higher (in the order of 4 to 5 m/s) in case of aortic stenosis, resulting in extremely elevated shear stresses believed to be responsible for increased risk of hemolisis [166], as well as elevated transvalvular pressure drop, with increase of ventricular afterload and increased risk for ventricular remodeling. On the other extreme of the spectrum, blood can stagnate in areas of the pathologically dilated ventricles, potentially leading to thrombus formation and increased risk of embolism [167]. Blood is a connective tissue with a particular rheological behavior, which makes it particularly important to account for what happens at the smaller spatial and temporal scales of hemodynamics.

Intra-cardiac hemodynamics is intrinsically a multi-physics problem, since the mechanical interaction of blood and myocardial walls is a major determinant of cardiac flow. *Fluid-structure interaction (FSI) models* are therefore well suited to describe the coupled system [168–176]. In a general formulation, such models describe the joint dynamics of two bodies (one solid, one fluid) that share an interface where they mutually exchange mechanical constraints. In the study of biological systems, a very popular model for blood dynamics is based on the Navier–Stokes equations for a Newtonian fluid [177]. As discussed in Section 1.3, myocardium can be modeled as a complex solid material with an active and a passive behavior; other parts of the heart such as valves and other connective tissues can be described as purely passive visco-elastic materials. The solid is free to move in space under the mechanical action of the fluid, as well as other constraints and according to its active behavior. The fluid changes its configuration according to its boundary conditions, which include the motion of the fluid-solid interface. At the fluid-solid interface, blood moves together with the structures it touches (no-slip condition). Mechanical stresses are transferred seamlessly across the interface.

In this section we review commonly used modeling approaches for the detailed study of intra-cardiac hemodynamics, including the mechanical coupling of blood with the surrounding tissues.

1.4.2.1 Modeling intra-cardiac blood flow

The Navier–Stokes equations governing the dynamics of a fluid read:

$$\rho \frac{\partial \mathbf{u}}{\partial t} + \mathbf{u} \cdot \nabla \mathbf{u} = \mathrm{div}\mathrm{T} + \mathbf{b}, \tag{1.21}$$

$$\mathrm{div}\, \mathbf{u} = 0, \tag{1.22}$$

where $\mathbf{u}$ is the fluid velocity, ρ its density, T is the stress tensor and $\mathbf{b}$ is a vector field of bodily forces. The first equation expresses the conservation of momentum, in the form of the balance between the acceleration of the fluid on the left hand side and the total sum of the forces on the right hand side. The second equation expresses the conservation of mass for the case of an incompressible fluid.

In a Newtonian fluid, the stress tensor takes the form:

$$\mathrm{T} = 2\mu \mathrm{D}(\mathbf{u}) - \nabla p, \tag{1.23}$$

where μ is the fluid viscosity and p its pressure, while $\mathrm{D}(\mathbf{u})$ is the strain rate tensor defined as $\mathrm{D}(\mathbf{u}) = \frac{\nabla \mathbf{u} + \nabla \mathbf{u}^T}{2}$. Substituting in Eq. (1.21) we obtain the formulation of the Navier–Stokes equations for Newtonian incompressible fluids:

$$\rho \frac{\partial \mathbf{u}}{\partial t} + \mathbf{u} \cdot \nabla \mathbf{u} - \mu \Delta \mathbf{u} + \nabla p = \mathbf{b}, \tag{1.24}$$

$$\operatorname{div} \mathbf{u} = 0. \tag{1.25}$$

This is a system of equations in variables $\mathbf{u}$ and p, valid in a portion of the 3D space $\Omega \in \mathbb{R}^3$, with proper initial and boundary conditions.

As mentioned above, in a fluid-structure interaction problem the interface between fluid and solid is free to move subject to the forces exchanged by the two bodies. This means that at least part of the boundary $\Gamma_w \in \partial\Omega$ changes position over time, resulting in a deformation of Ω. This is naturally described in the elastodynamics problem, when Γ_w is considered as part of the solid body and its motion is modeled as the motion of all other material points in the solid. The Navier–Stokes equations are instead usually posed in an Eulerian frame, so that the fluid is observed as it flows through a fixed region of space. To handle the deformation of the fluid domain Ω, different approaches can be followed depending on the method used to discretize the equations. For instance, in the case of finite elements methods, care has to be taken so that the motion of the domain does not cause degeneration of the volumetric mesh used to discretize it. Mesh update can be performed using smooth operators, that extend the velocity of Γ_w to the entire fluid domain. In case of large displacements, grid distortion may be unavoidable: in which case the mesh may have to be re-computed adding computational complexity to the method. Methods relying on fixed fluid grids such as the immersed boundary and fictitious domain methods remove the complexity associated with continuous updates to the computational domain, however tend to suffer from larger approximation errors in the neighborhood of Γ_w due to interpolation effects. More details and comments on numerical discretization techniques are provided in Section 2.4.

Domain Ω can change substantially over the heart cycle, resulting in volume variations in the order of 50% of the maximum value in normal physiological conditions [137]. In an average male subject, the resulting aortic flow rate is in the order of 5 l/min, and with most of the ejection happening during the systolic phase, this leads to peak flow rate values in the order of 500 ml/s. Peak velocity values measured in the ascending aorta can reach as high as 9 m/s [178] in systole, while being negligible in diastole. The fast, pulsatile dynamics of blood flow can induce flow instabilities and transient turbulence effects [101,179], which need to be accounted for when designing the discretization approach. Flow stability is often characterized with a single parameter, the Reynolds number, representing the ratio of intertial forces to viscous forces. The Reynolds number is proportional to the mean flow velocity and to the characteristic length of the domain, and inversely proportional to the fluid viscosity. Large values of the Reynolds number identify scenarios in which inertial forces are dominant over viscous forces, and vice versa. Typical values in the human cardiovascular system range from few thousands (in bigger arteries such as aorta, iliac arteries, brachial arteries and in bigger veins) to less than 1000 in medium-size vessels (such as carotid arteries, the main coronary arteries and medium-size veins) and even less than 1 (arterioles, capillaries, venules). At Reynolds number values around 2300, a steady flow could become turbulent; pulsatility in blood flow however causes the transition to turbulence to appear at higher values of the Reynolds number, to then disappear after the deceleration phase. The critical Reynolds number depends in general on the rate of change of velocity, as well as on the geometry of the domain [101]. Turbulence in blood flow is associated with oscillating pressure and increased wall shear stresses on the boundary of the domain. These effects are particularly relevant when studying non-physiological situations that can increase the likelihood of flow instabilities, as in the case of valvular disease [180] or the presence of devices such as artificial valves [181] or ventricular assist devices [182], and in which an altered hemodynamics environment could promote disease progression.

1.4.2.2 Fluid-structure interaction

A fluid-structure interaction cardiac computation system integrates the electrophysiology-controlled biomechanical components with 3D computations of intra-ventricular flow dynamics, as well as valve kinematics or dynamics. The FSI problem can be solved with a monolithic approach or a partitioned approach. In a monolithic approach, the equations governing the fluid dynamics and the elastodynamics are solved simultaneously, therefore implicitly accounting for the interface coupling conditions. This approach typically requires the implementation of ad-hoc solvers, and can be computationally demanding due to the sheer size of the resulting system of equations. On the other hand, it generally allows a more stable solution of the coupled problem. Partitioned approaches have the advantage of allowing the use of off-the-shelf solvers tailored to the individual problems (fluid dynamics and elastodynamics). However, fulfilling the coupling conditions may require iterative solution of the problem at each time step, potentially resulting in significant computational overhead.

We focus our description on a *partitioned* approach. An illustration of possible components for such an FSI computation system is given in Fig. 1.13. We note the inclusion of 3D valve geometry information, as relatively accurate models of the valves (especially atrio-ventricular ones) are necessary for reproducing many of the key features of the intraventricular flow field [183].

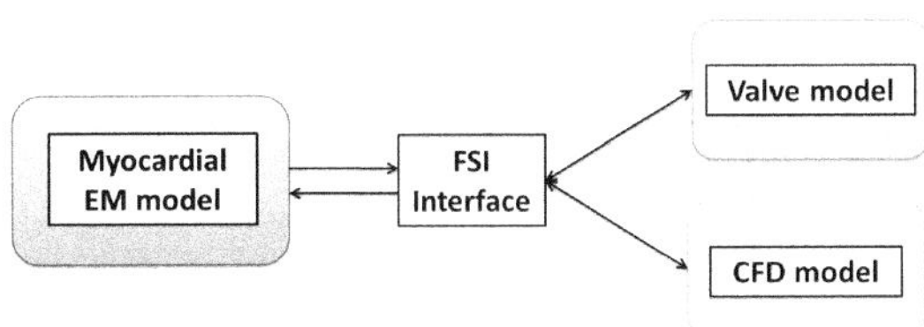

Figure 1.13. Fluid structure interaction system for cardiac haemodynamics computation. The interactions between the electromechanical model, valves and the computational fluid dynamics (CFD) model are controlled by the FSI interface module.

An FSI module is, at its core, a coupling interface between the myocardial electromechanical (EM) solver, the valves and the 3D computational fluid dynamics (CFD) solver, and controls the proper exchange of information. Its usefulness is maximized when one needs to integrate independent computational modules that do not share data at a "deep" level, and need a special handler of information between them. At each time step, the cardiac FSI module would need to enable the following interactions:

- sending stress load information from fluid to the biomechanics and valve modules
- sending endocardial wall and valve positions and velocities to the CFD solver
- exchanging biomechanical and valve models stress and anatomic constraint information.

In Chapter 2 we will look in detail at a possible realization of a cardiac FSI system. We have discussed and described so far its two main components: the biomechanical model and the 3D CFD solver. We end this part with a closer look at the third component, the valve model.

Valve models and FSI

Valves, especially the atrio-ventricular ones (and any of them when pathology is present) can introduce large flow disturbances that need to be taken into account in certain circumstances. For example, a proper integration of the mitral valve would induce a posterior flow deflection, and create a natural circulation pattern in the ventricle [184], which may not happen if the valve is not correctly modeled. For modeling the valves one can distinguish three approaches:

- **dynamic**: modeling the valve as a deformable structure and computing its motion using FSI
- **kinematic**: using a kinematic model of the valve, with the geometry and movement of the leaflets prescribed from previous measurements
- **kineto-dynamic**: combining the above two approaches.

The dynamic approach [88–95] requires a significant modeling effort for the mitral valve, so that the constitutive and structural properties of such a complex structure can be identified properly. On one hand, with an eye focused on clinical applications, we are still far from being able to deploy patient-specific biomechanical mitral valve models. On another hand, FSI models have shown their usefulness in the study of natural and prosthetic aortic valves and also prosthetic mitral valves. This is because the structure and kinematics of mechanical prosthetic valves is much simpler than the real counterparts. Two recent review articles [96] and [97] provide more extended summaries of the current status on the patient-specific simulation of cardiac valves.

The kinematic approach uses pre-defined mesh sequences of the (usually) mitral mesh opening-closing cycle, generated from imaging sequences (usually CT or US, but MRI can also be used with extra modeling effort). This approach is not useful if the interest is to predict the motion of the leaflets themselves, but in certain clinical applications the kinematic model of the mitral valve could be a viable approach. An example is LV thrombosis, where the interest lies in the flow patterns in the left ventricle. Examples of using the kinematic approach can be found in [131], where flow patterns are analyzed in presence of pathology. This can be considered a 1-way FSI system, where the valve motion does not respond to stress/pressure gradients.

The third approach combines the two previous types, and was also used for some of the results reported in Chapter 2. In [98] the authors proposed a reduced degree of freedom model of the mitral valve, in which the opening angle is determined dynamically by matching an interface condition. In the approach presented here, the dynamics is handled by a reduced degree of freedom system, namely the pressure-driven 0D valve system already presented in section 1.4.1. This dynamic system then maps its opening phase to the pre-computed matching kinematic frame, which is used to impose no-slip boundary conditions on the surrounding fluid. In Chapter 2 we provide more details about this approach.

1.5 Current approaches to parameter estimation

As computational models mature and numerical solvers become more efficient, scientists started to investigate how these models could be applied for clinical problems [5,6,8,9,47]. Several personalization approaches have been proposed, from entirely manual to automatic image-based methods. For instance, a common arterial Windkessel (WK) model personalization technique can be found in [185,186]. However, personalizing electromechanical models is a more challenging endeavor due to the significantly higher computational cost of the forward simulations and the larger number of free parameters to estimate. Several categories of methods can be identified in the literature, including gradient-based and gradient-free inverse optimization methods, data assimilation methods, methods based on machine learning (ML), and stochastic approaches. The following sections provide a brief overview of each of these strategies. One specific implementation is detailed in Section 2.5.

1.5.1 Inverse optimization

The standard approach to estimate tissue parameters from data is based on optimization, as in [6,18] for instance. The idea is to design a cost function that calculates the distance between computed parameters and their clinical measurements, and minimize that cost function by tuning the free parameters of the model. A common choice is to use gradient-free methods [187], as the cost functions and their derivatives with respect to model parameters can be complex to derive analytically. For instance, in [188], the authors proposed a gradient-free optimization method to estimate patient-specific biomechanical contractility from myocardial velocity. Mathematical representation of heart shape and motion was also exploited in [189] to evaluate the goodness of fit between the biomechanical simulation and the real heart. A multi-step optimization procedure was proposed in [190] to estimate left ventricular passive myocardial properties. More recently, a combination of one-dimensional parameter sweeps with gradient-free optimization was presented in [191] to estimate myocardial passive material parameters on a single-ventricle model. A similar multi-step idea was presented in [192], where parameter sweeps (exhaustive search) are performed iteratively on pre-defined one- and two-dimensional sample grids. Unfortunately, methods based on parameter sweeps quickly become intractable as the number of parameters to optimize grows. They can therefore enable the exploration of the full landscape of the objective function only at a pre-defined, usually coarse resolution. Hierarchical, multi-level approaches have also been explored to cope with the curse of dimensionality [18,193], as well as multi-model techniques [194]. Finally, the authors in [195] presented an approach for iterative regional personalization of cardiac electrophysiology models, tailored for patients with left-bundle-branch block.

1.5.2 Data assimilation

Data assimilation approaches frame the personalization problem in a way that aims at identifying unknown variables using observations of a dynamical system. At each iteration of the personalization loop, a forecast of the computational model and the assimilated observations are compared to estimate the current goodness of fit between model and data, considering uncertainty in the state and in the observations. Two categories of data assimilation methods can be distinguished: variational [196] and filtering approaches [121,197]. The latter includes the Kalman filtering approach and its extensions (e.g. unscented Kalman filters). Filtering approaches have been used for global [198] as well as regional [121] active contractility estimation, or passive material parameter estimation [192].

1.5.3 Machine learning

The past few years saw the development of data-driven machine learning (ML) methods, such as the one-shot global EP personalization approach described in Section 5.2.1. In another work [199], the authors learn the non-linear relationship between heart motion, derived from temporal images, and parameters of electrical propagation. These examples demonstrated that machine learning approaches can learn the non-linear mappings from observations to cardiac model parameters. Furthermore, carefully generating synthetic training databases using computational models can enable effective training schemes even when real data is scarce. Genetic algorithms have also been investigated as ways to estimate model parameters from clinical data [129, 200]. In particular, such algorithms were combined with a multi-fidelity modeling approach, where a detailed 3D model was approximated by an efficient 0D surrogate model to speed up the overall estimation process of active and passive tissue parameters [129].

1.5.4 Stochastic methods

Finally, stochastic methods aim to find the "optimal" parameter set along with their confidence values, considering uncertainties in data and models [26,82,201].

1.5.5 Streamlined whole-heart personalization

Comprehensive and fast 3D whole-heart model personalization is nowadays becoming possible. In [18], the authors presented a first approach. However, it involved significant manual steps and its robustness to different patients, pathologies and data quality remains an open question. To our knowledge, only few comprehensive, integrated pipelines have been presented so far [6,9] to personalize anatomy, electrophysiology, biomechanics and hemodynamics in a streamlined, consistent and automatic fashion. Section 2.5 will present in details one such framework.

1.6 Summary

This chapter introduced models and approaches for multi-physics, multi-scale modeling of heart function. Several modeling options are available, going from very detailed models to coarse ones. Like in other scientific domains, it is likely that one model would not fit all the clinical questions that could be explored and potentially answered using multi-physics approaches. It is therefore the task of the researchers, modelers, computer scientists and physicians together, to identify 1) the right level of model details to use, 2) what computational framework and numerical approximations to afford and 3) which parameters should be estimated for personalized simulations. One strategy that proved useful is to start from the clinical challenge to address, and answer these questions accordingly. This was the approach used for all examples described in this book, resulting in a wide variety of models, from simple to complex, but all adjusted to the clinical workflow that was considered. The next chapter presents implementation strategies for patient-specific modeling from clinical data, the basis of the more advanced solutions based on AI presented in the second part of the book.

2

Implementation of a patient-specific cardiac model

Viorel Mihalef, Tommaso Mansi, Saikiran Rapaka, Tiziano Passerini
Siemens Healthineers, Princeton, NJ, United States

Creating a patient-specific model of the heart starts with a set of medical images, like for instance CT or MRI, to estimate a detailed model of the patient's heart anatomy. For bi-ventricular simulations, the resulting anatomical model comprises the geometry of the two ventricles, the orientation of myocardial fibers, and other spatially-varying information such as location of scarred tissue. For detailed computations of blood flow one could also include models of the valves, which can be provided as 3D anatomies integrated geometrically with the myocardium, or as reduced models, with 0D functional representations. Atrial models should be included either as 3D geometric models, if needed for whole-heart computations (e.g. atrial electrical signal propagation or 3D atrial hemodynamics), or as reduced (0D) models. The anatomical model is used as computational domain for three solvers: cardiac electrophysiology, tissue biomechanics, and hemodynamics (intracardiac flow as well as flow in the arteries, the atria and the veins, applied as boundary conditions to the electro-mechanical model). Finally, clinical parameters are computed from the simulated cardiac dynamics (e.g. stroke volume, ejection fraction, etc.). These parameters are used as targets during model personalization, but also as ways to assess the effect of the simulation scenario (baseline or therapy) to investigate. Fig. 2.1 illustrates the main blocks of the model presented in this chapter. Every component is designed independently, with data flowing between each component as illustrated by the black arrows.

For each component, various simplifying assumptions can be made in order to allow for efficient personalization and computation. Simplification decisions are often made based on the clinical application. For instance, studying the effect of conduction delays may be achieved with a graph-based model of cardiac electrophysiology, whereas complex arrhythmias demand a more complex, phenomenological if not biophysical model. Conversely,

Artificial Intelligence for Computational Modeling of the Heart
https://doi.org/10.1016/B978-0-12-817594-1.00012-7

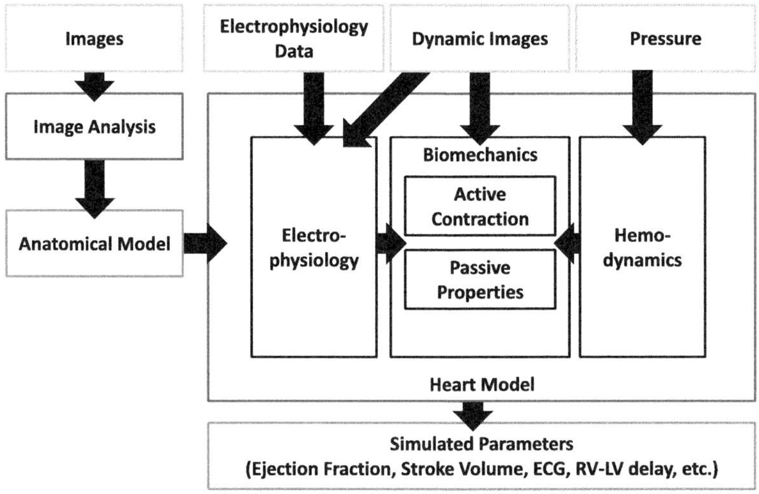

Figure 2.1. Elements, with input and output, of a typical computational model of the heart. Green (mid gray in print version): input data. Red (dark gray in print version): processing units. Orange (light gray in print version): output data. Arrows denote data flow.

mechano-electrical feedback (modification of electrical conductivity due to mechanical pathologies) could also be modeled if required by the application. This chapter provides implementation details of each component, knowledge that will be needed for the remainder of the book.

2.1 Anatomical modeling

A typical anatomical modeling pipeline starts by extracting the cardiac surfaces from medical images, followed by the estimation of a volumetric mesh, automatically annotated with various anatomical, physiological and functional properties. An illustration of this pipeline is given in Fig. 2.2. The following sections describe each step of the anatomical modeling process in more details.

2.1.1 Medical image segmentation

The first step consists in segmenting the medical images to extract the different cardiac structures to be modeled. Several methods exist, from completely manual and interactive to automated approaches powered by artificial intelligence (AI). Chapter 3 describes recent AI-based methods that achieve high level

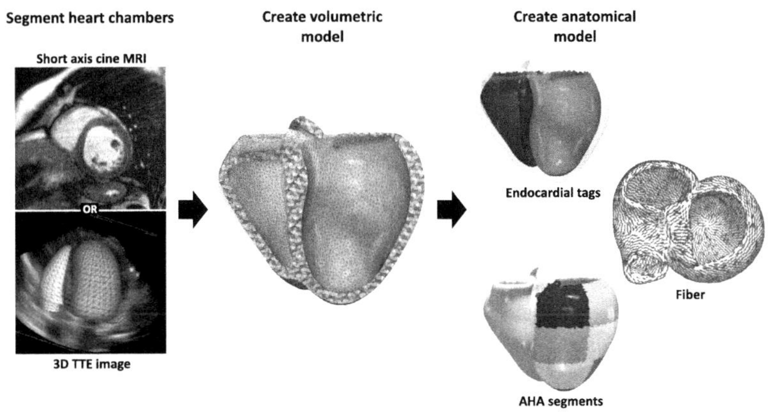

Figure 2.2. Overview of the anatomical modeling pipeline based on medical image segmentation.

of accuracy and reproducibility, two necessary requirements for clinical applications. Traditionally, methods based on machine learning and shape models have been employed [31,32]. For instance, in [202], the authors use a machine-learning framework to segment cardiac structures. Marginal space learning [31] was introduced as an efficient way to learn high dimensional models and perform fast online search by operating on spaces of increasing dimensionality. In practice, the location, the orientation, and the scale of the structure of interest are identified sequentially. Anatomical landmarks are then detected inside the resulting region of interest, and a shape model is fitted to match the cardiac surfaces. The model is further refined using boundary detectors. At each step, detectors trained on large databases of annotated images are used. In particular, probabilistic boosting trees [203] are employed, as they can account for patterns of large intra-class variability for complex data distributions. Fig. 2.3 illustrates some segmentation results on various imaging modalities.

One advantage of using shape models is their inherent parameterization. Anatomical landmarks are explicitly encoded: semantic associations to the underlying anatomy across patients and modalities is thus automatically provided (Fig. 2.4). As a result, these models are highly modular. For example, cardiac resynchronization therapy (CRT) in-silico analysis would benefit from having the models of both left (LV) and right (RV) ventricles, while advanced flow computation applications would benefit from inclusion of valve models along with all chambers. In general, such models can provide explicit geometric representations for the left ventricle endocardium, epicardium, mitral annulus, left ventric-

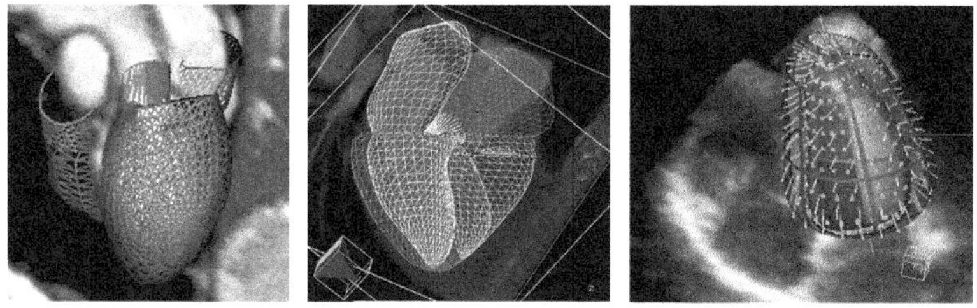

(A) (B) (C)

Figure 2.3. From left to right: final geometrical models extracted from computed tomography (CT), magnetic resonance image (MRI) or ultrasound data.

ular outflow tract, ventricular regions, tricuspid and pulmonary valve locations. The right side of Fig. 2.4 shows a similar degree of detail for the aortic (top) and mitral (bottom) valve models, as proposed by [32]. Temporally consistent segmentation of the chordae tendinae and ventricular trabeculation is still in its infancy, but it will have its place in the whole framework once it matures.

2.1.2 Meshing and tagging

Once the cardiac structures are segmented from medical images, a volumetric mesh model is defined as the computational domain. To simulate ventricular function for instance, the LV and RV surfaces segmented from the images are automatically fused together into a single mesh representing the thick myocardium. If the RV epicardium is not visible, simple geometric extrusion with a user-defined thickness can be used to model it. The mesh is then tetrahedralized: namely filled with tetrahedra elements, the spatial discretization used in this chapter. Other types of volumetric elements could also be used (e.g. hexahedral elements). However,

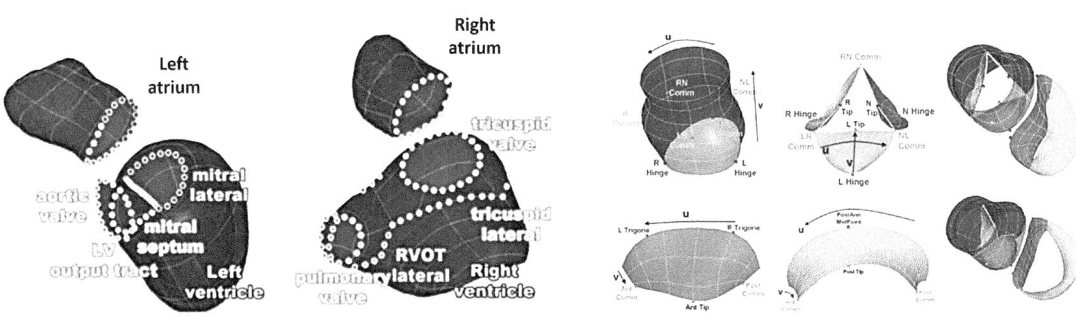

Figure 2.4. Ventricular models (left images) and valvular models (right images) are parameterized and tagged.

the meshing techniques would become more involved, in particular when the underlying shape is as complex as a patient-specific, diseased heart. The anatomical structures are automatically annotated on the tetrahedral mesh by leveraging the parameterization of the segmented surfaces (Fig. 2.5). Myocardium scar or fibrosis, identified in images like delayed-enhancement MRI, are also mapped onto the anatomical model to account for their impact on cardiac function.

Figure 2.5. Tagged surface meshes (left and middle panels) and fused tetrahedral mesh (right panel).

The fibrous, collageneous tissue is also tagged on the model, with the aim to properly represent its inactive electro-mechanical properties. Based on literature reports, a rule-based classification of fibrous tissue can be designed. This comprises the fibrous rings of the pulmonary and aortic valves, as well as fibrous connections linking these rings to the atrioventricular valves [204,205]. We therefore tag as fibrous tissue all mesh elements in the left ventricular outflow tract as well as all mesh elements in the right ventricular outflow tract and above the plane of the atrioventricular valves (Fig. 2.5). Finally, for regional parameter estimation or function analysis, a subdivision in segments (according to the definition proposed in [206]) is automatically defined based on the tags, as illustrated in Fig. 2.6.

Figure 2.6. Automatic subdivision of the biventricular anatomical model according the segment definition in [206]. A 17-segment model is represented for two patient-specific geometries.

2.1.3 Computational model of the cardiac fiber architecture

The implementation described in this chapter relies on a rule-based model of myocardium architecture derived from ex-vivo studies [49,55,56], following a similar strategy as in [45, 57]. The model, which covers fiber orientation and fiber sheets, can be adapted to the investigated pathology. Let $\xi = (\mathbf{f}, \mathbf{s}, \mathbf{n})$ be the local, orthonormal fiber coordinate system, where $\mathbf{f}$ is the fiber direction, $\mathbf{s}$ the sheet direction and $\mathbf{n}$ the sheet normals. Fiber and fiber sheet orientations are determined by the angles α and β defined with respect to the circumferential ($\mathbf{e}_0$), longitudinal ($\mathbf{e}_1$) and radial ($\mathbf{e}_2$) axes (Fig. 2.7). The elevation angle α of the fibers, i.e. their angle with respect to the short axis plane, varies linearly across the myocardium, from $\alpha = -70$ degrees on the epicardium to $\alpha = 0$ at mid-wall to $\alpha = +70$ degrees on the endocardium. These values could change under disease conditions, like for instance in hypertrophic cardiomyopathy or around an area of myocardium infarction. Fiber sheet organization is less marked and varies more across individuals. The angles are traditionally assumed to range between $\beta = +45$ on the epicardium to $\beta = -45$ on the endocardium, as measured in [56].

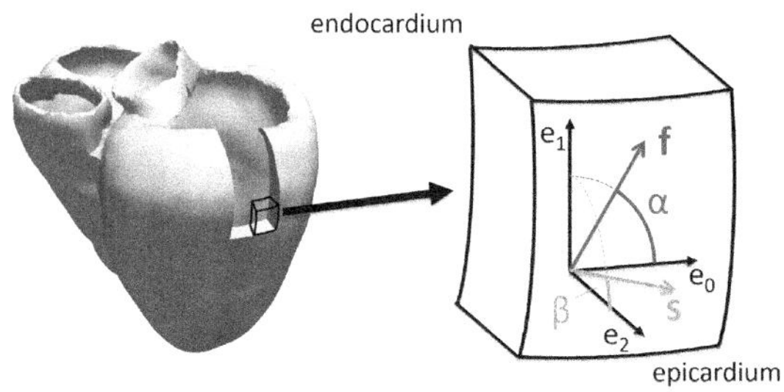

Figure 2.7. Fibers and sheets orientation from endocardium to epicardium. A local coordinate system based on the circumferential ($\mathbf{e}_0$), longitudinal ($\mathbf{e}_1$) and radial ($\mathbf{e}_2$) directions is used to define a rule-based model of fiber orientations.

The construction of the model is done in two steps. First, the fiber coordinate system ξ is estimated from the apex to the base plane, where endocardial and epicardial fibers and fiber sheets have a constant angle (Fig. 2.8, left panel). The angles α and β are then computed throughout the myocardium using linear interpolation. In a second stage, the fiber coordinate system is calculated from the base plane to the valves. Special care is taken in order to achieve smooth and realistic fiber variations. The coordinate system ξ is first fixed around the valves: fibers are cir-

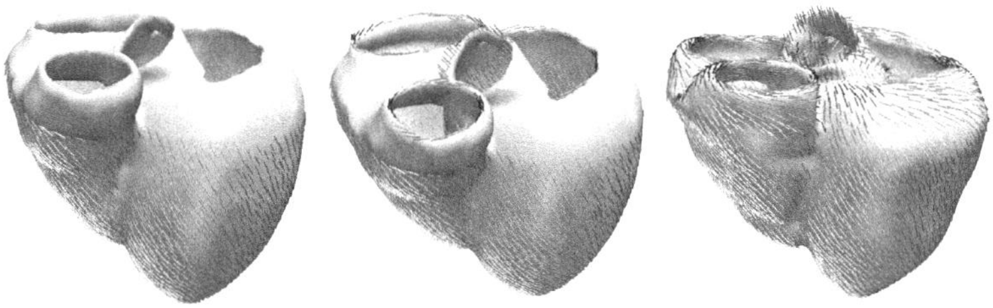

Figure 2.8. Fiber estimation processing. *Left panel*: apex to base fiber estimation using a rule-based model. *Mid panel*: prescription of fiber orientation around the valve. *Right panel*: geodesic interpolation of fibers from the base to the valves.

Figure 2.9. Example of computed fibers and fiber sheets on a patient-specific anatomy.

cumferential around the mitral, tricuspid and pulmonary valves and longitudinal around the left ventricle outflow tract (Fig. 2.8, mid panel); sheet normals are oriented towards the center of valves. We then perform a geodesic interpolation of the coordinate system ξ on the endocardium and epicardium surfaces between the base plane and the valves (Fig. 2.8, right panel). The Log-Euclidean framework [207] is used to ensure the orthonormality of ξ during the interpolation. The resulting surface coordinate system is finally interpolated throughout the myocardium using the same interpolation scheme. Fig. 2.9 illustrates the final patient-specific myocardium architecture. It is worth noting that all the above operations are performed node-wise. However, a cell-based definition may be obtained from the nodal values using barycentric interpolation in the Log-Euclidean space. This could be useful depending on the numerical scheme adopted for the discretization of the differential equations governing the physiological model.

2.1.4 Torso modeling

Patient-specific torso geometry can be either segmented automatically from images [202] or, when the torso is not fully visible in the images, estimated from an atlas by manually registering the model to the patient. If the images do not have the sufficient field of view to align the torso, or are not at all available such as in the case of ultrasound-based workflows, the relative position of the heart in the thoracic cavity is used as reference for the alignment. Fig. 2.10 illustrates a torso model with superimposed virtual ECG electrodes.

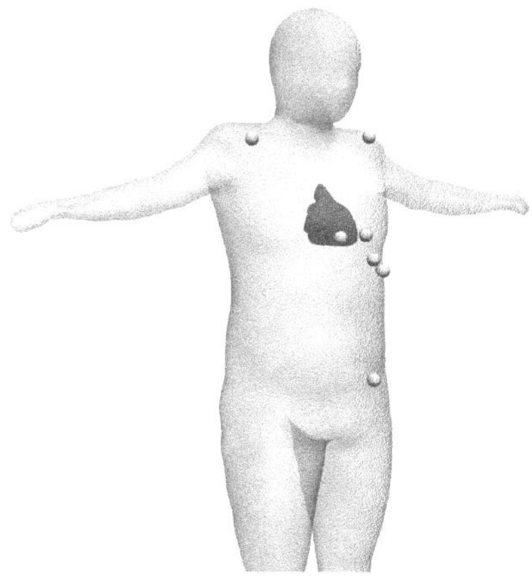

Figure 2.10. Image of the torso avatar used for fitting the imaging data, with the standard 12-lead ECG leads in place.

2.2 Electrophysiology modeling

As discussed in the previous chapter, electrophysiology is an intrinsically multi-scale phenomenon. The depolarization of a myocyte takes few microseconds, compared to a typical one-second heart-cycle, corresponding to a factor of 10^6 between timescales. Spatially, the depolarization wave propagates as a sharp front with a length-scale of a few microns compared to the size of the heart (in the order of 10–15 cm), corresponding to another factor of 10^4–10^5 in spatial length scales [208]. Depending on the problem of interest or the desired model fidelity, different modeling choices can be made and the computational methods need to be properly designed to capture the large differences in scales in the phenomena described by the models. The problem is further complicated by the complex geometry of the heart and the heterogeneous and anisotropic nature of the myocardial tissue. The following sections describe in more details some possible modeling choices, discussing their advantages and potential implementations.

2.2.1 LBM-EP: efficient solver for the monodomain problem

Numerically, cardiac electrophysiology (EP) models are usually solved using finite difference methods (FDM) or finite element methods (FEM) [209]. In FDM, the computational domain is spatially discretized into a Cartesian grid, and the spatial derivatives occurring in the differential equation are approximated using differences over neighboring nodes [209]. These methods tend to be easier to implement, and have the advantage of speed thanks to efficient implementations based on the structured connectivity of the grid nodes. However, the boundary conditions can be difficult to impose over complex geometries. Furthermore, Clayton et al. [209] showed that the implementation of anisotropic conductivity can be difficult, with the propagation speed depending significantly on the fiber orientation with respect to the Cartesian grid axes.

Finite element methods [210,211] on the other hand solve a weak formulation of the governing equation derived by the Galerkin approach. The computational domain is divided into a set of simple elements, and the weak formulation is solved using (typically) low order polynomial interpolation inside the elements. This method has the advantage of being able to adapt to complex geometries easily. The implicit and semi-implicit formulations typically used for time discretization require the solution of a linear system of equations at every timestep, and can be computationally slow. For this reason, explicit time stepping schemes using lumped mass matrices have become popular, enabling fast computations [212]. However, the solution accuracy is sensitive to the way in which mass-lumping has been performed [213].

Recently, the Lattice-Boltzmann Method (LBM) has gained interest for solving complex reaction–diffusion–advection equations, in particular the Navier–Stokes equation in Computational Fluid Dynamics. This model has had great success in the fluid dynamics community over the last couple of decades (see [214,215] for comprehensive reviews), and has also been applied to other problems with a reaction–diffusion nature [216]. Complex geometries can be handled easily using LBM by level-set-based extrapolation techniques, enabling second-order accurate boundary conditions. Furthermore, LBM depends on highly localized computations with minimal communication between neighboring nodes, which makes the algorithm inherently well suited for parallel architecture. Unlike the FDM which is also based on a Cartesian lattice, the spatial derivatives are not explicitly computed but are obtained implicitly as part of the algorithm.

The Lattice-Boltzmann method was recently proposed [217] as an alternate approach for fast solution of general, monodomain electrophysiology models (the LBM-EP method), and was applied to real patient anatomies (Fig. 2.11). Its implementation on Graphical Processing Units (GPUs) showed nearly real-time performance on a regular workstation with off-the-shelf GPUs [218,219]. This computational performance enabled fast patient-specific calibration of a cardiac electrophysiology model from images and 12-lead ECG signals [220] and this model was successfully applied on 19 real patients with Dilated Cardiomyopathy (DCM).

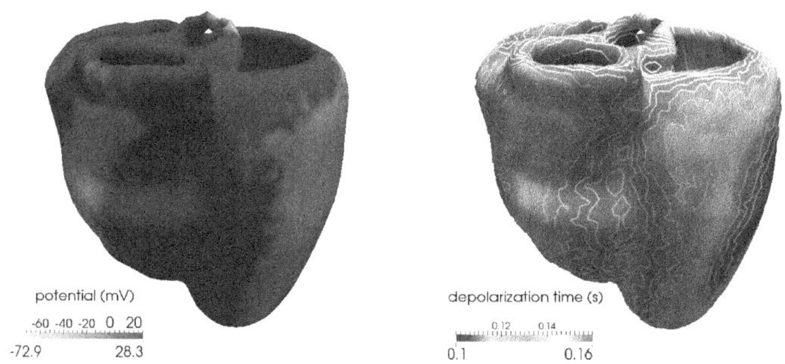

Figure 2.11. (Left) Snapshot of the activation potential propagation through the myocardial tissue and (right) the resulting map of the activation times, computed by solving the monodomain model of tissue electrophysiology with M-S cellular model and LBM computational method.

In contrast to the finite element and finite difference formulations, which solve the governing equations for the action potential, the lattice-Boltzmann method solves a more fundamental kinetic equation, i.e., the Boltzmann equation. This method has its origins in the approaches based on cellular automata, and tries to utilize the simplicity of the physics at the microscopic scale to produce emergent macroscopic behavior. The fundamental kinetic variable is the distribution function $f(\mathbf{x}, \mathbf{e}, t)$, which gives the probability of finding a particle at location $(\mathbf{x}, t)$ traveling with a velocity $\mathbf{e}$. The lattice Boltzmann method is a special discretization of the kinetic model in the velocity space, where the particle velocities are restricted to belong to a small set of discrete velocities $\mathbf{e}_i = \{\mathbf{e}_1, \mathbf{e}_2, \ldots, \mathbf{e}_N\}$. This results in a vector of distribution functions $\mathbf{f} = \{f_1, f_2, \ldots, f_N\}$, where $f_i(\mathbf{x}, t) = f(\mathbf{x}, \mathbf{e}_i, t)$, and each component is strictly a function of only space and time. The velocity vectors $\mathbf{e}_i$ are chosen to be the links connecting every lattice node to its 6 nearest neighbors on the Cartesian grid, as well as a link to itself. The trans-membrane potential is related to the distribution function through $v = \sum_i f_i$. The dynamics of the distribution function evolves through the following linearized Boltzmann equation:

$$\frac{\partial f_i}{\partial t} + \mathbf{e}_i \cdot \nabla f_i = \Omega_{ij}(f_j - f_j^{(0)}) + s_i, \tag{2.1}$$

where Ω_{ij} is a matrix which operates on the deviation of the distribution function from its equilibrium value $f_i^{(0)} = v/7$ and s_i is the contribution of any sources (ionic and stimulus currents). The analysis is significantly easier operating in a space of moments of the distribution function, instead of directly operating over **f**. For this reason, we introduce an orthogonal transformation matrix M which transforms the distribution function into a set of linearly independent moments. For the 7-velocity lattice introduced above, the matrix M is chosen to be:

$$\mathrm{M} = \begin{bmatrix} 1 & 1 & 1 & 1 & 1 & 1 & 1 \\ 1 & -1 & 0 & 0 & 0 & 0 & 0 \\ 0 & 0 & 1 & -1 & 0 & 0 & 0 \\ 0 & 0 & 0 & 0 & 1 & -1 & 0 \\ 1 & 1 & 1 & 1 & 1 & 1 & -6 \\ 1 & 1 & -1 & -1 & 0 & 0 & 0 \\ 1 & 1 & 1 & 1 & -2 & -2 & 0 \end{bmatrix}. \tag{2.2}$$

The moments corresponding to the equilibrium distribution function can be seen to be $\mathbf{m}^{(\mathbf{0})} = \mathrm{M}\mathbf{f}^{(\mathbf{0})} = [v, 0, 0, 0, 0, 0, 0]^T$. Correspondingly, the moment vector $\mathbf{m} = \mathrm{M}\mathbf{f}$ contains local information related to the potential and its higher derivatives. The first component, $m_0 = v$ is the transmembrane potential itself, and the next three moments can be shown to be related to the components of its gradient, $(\partial v/\partial x_1, \partial v/\partial x_2, \partial v/\partial x_3)$. The last three moments are related to higher order derivatives, which don't influence the solution up to second order accuracy in spatial resolution. The operator, $\boldsymbol{\Omega}$ can correspondingly be rewritten as a transformation to the moment space, followed by the relaxation in moment space and then transform back into distribution function space, i.e., $\boldsymbol{\Omega} = \mathrm{M}^{-1}\mathrm{SM}$.

Discretizing Eq. (2.1), we get:

$$\mathbf{f}(\mathbf{x} + c\mathbf{e}_i\delta t, t + \delta t) = \mathbf{f}(\mathbf{x}, t) - \mathrm{M}^{-1}\mathrm{SM}(\mathbf{f} - \mathbf{f}^{(\mathbf{0})}) + \delta t\mathbf{s}(\mathbf{x}, t), \tag{2.3}$$

where $c = \delta x/\delta t$ is the speed of the particles in lattice units. This process is usually represented as two distinct steps: collision and streaming. The collision steps updates the distribution function at each location to mimic the scattering of particles due to molecular collisions and due to applied external sources (currents). The post-collision distribution function is given as:

$$\mathbf{f}^*(\mathbf{x}, t) = \mathbf{f}(\mathbf{x}, t) - \mathrm{M}^{-1}\mathrm{SM}(\mathbf{f} - \mathbf{f}^{(0)}) + \delta t\mathbf{s}(\mathbf{x}, t). \quad (2.4)$$

The streaming step merely propagates each component of the distribution function towards the nearest node along its lattice direction

$$f_i(\mathbf{x} + c\mathbf{e}_i\delta t, t + \delta t) = f_i^*(\mathbf{x}, t). \quad (2.5)$$

This process is depicted in the schematic in Fig. 2.12. It can be shown that with the proper definition of S the dynamics of the distribution functions tends asymptotically to the solution of the partial differential equations of the monodomain model.

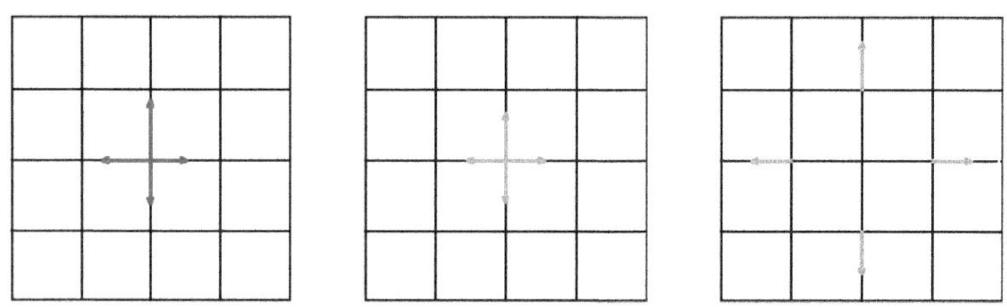

Figure 2.12. Description of the LBM-EP algorithm on a 2-D slice. The first image shows the pre-collision distribution in a node at the start of the step. The collision step redistributes the distribution function values (middle figure) and finally the post-collision values stream to the corresponding neighbors.

LBM-EP evaluation

Computational models of electrophysiology are extremely hard to validate, owing to the lack of analytical solutions. To this end, recently a benchmark study was conducted [221] to compare the predictions of eleven existing computational solvers. To eliminate geometric complexity, the problem domain was chosen to be a cuboidal piece of tissue of size $20 \times 7 \times 3$ mm. All the solvers were required to use the epicardial variant of the ten Tusscher and Panfilov cell model with prescribed model parameters. The computations were required to be done at prescribed spatial resolutions of 0.5 mm, 0.2 mm and 0.1 mm and the computations at each spatial resolution were run at three different time-steps of

0.005 ms, 0.01 ms and 0.05 ms. The tissue sample was stimulated in a cube of size 1.5 × 1.5 × 1.5 mm at one of the corners and the activation times were reported at all the corners, the center of the sample as well as a plane passing through the center. The problem geometry is shown in Fig. 2.13.

It was found that despite the simplicity of the problem geometry, there was considerable variability among the results, especially at the coarser grid sizes. There was inter-model variability, even if the underlying numerical scheme was the same (i.e., finite-element or finite-difference) due to differences in the nature of interpolations and mass lumping [213].

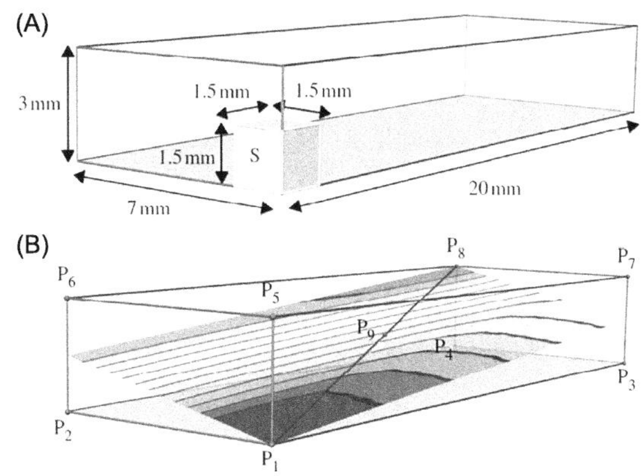

Figure 2.13. Schematic of the problem geometry from [221]. A tissue sample of size 20 mm × 7 mm × 3 mm is stimulated in the cube marked S. The activation times are reported at points P1 to P9, as well as on the slice shown in (B). (*Source: [221].*)

Since the authors have made available the detailed results for this benchmark, it could be used to demonstrate the validity of the LBM based approach described here. Fig. 2.14 shows the depolarization times along the diagonal P1–P8 of the computational domain obtained by the different groups who took part in the benchmark (boxes A-K) along with the LBM-EP results (box L). It can be seen that the depolarization times computed by LBM-EP are consistent with those obtained with existing finite-difference and finite-element based solvers. Interestingly, the results are closest to those obtained with the finite-element solver CHASTE (box A), with the coarsest resolution over-predicting the speed of the depolarization wave. The other Cartesian-grid based solvers have large errors at the coarsest grid size, significantly underpredicting the speed of the depolarization wave. The differences in depolarization time with changing spatial resolution are much smaller for LBM-EP. For this benchmark problem, LBM-EP demonstrates accuracy comparable to the existing state-of-the-art solvers.

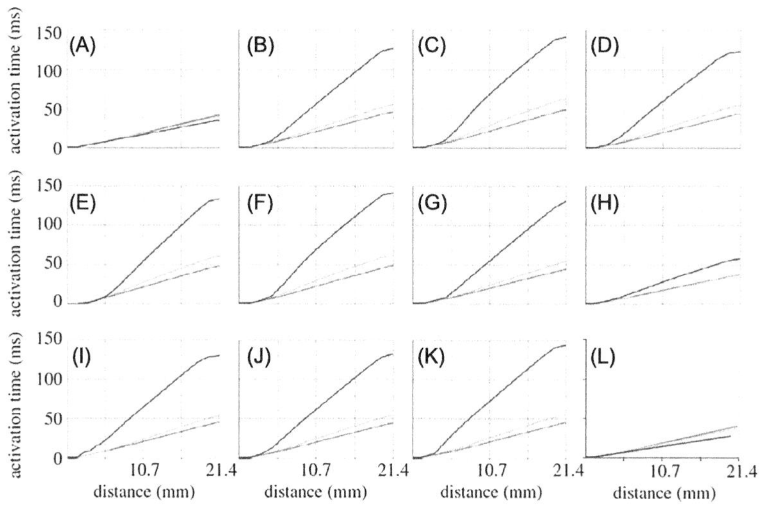

Figure 2.14. Comparison of the activation time computed from LBM-EP (box L) with those presented in [221]. Red (mid gray in print version), green (light gray in print version) and blue (dark gray in print version) lines represent solutions using a spatial resolution of 0.1 mm, 0.2 mm and 0.5 mm respectively. Codes A, B, C, E, F and H are finite element codes. Codes D, G, I, J, and K use finite differences.

This example was also used to demonstrate the computational performance of the LBM-EP algorithm. This exercise was run on a standard workstation with Xeon processor, 6 GB of RAM and a NVIDIA Quadro 4000 graphics card. Fig. 2.15 reports the number of lattice node updates (millions) per second (MLUPS) plotted against the number of computational nodes.

$$\text{MLUPS} = \frac{\text{Number of nodes} * \text{Number of time steps}}{\text{Total computational time}}$$

For each configuration, the execution time for the finest time-step ($\delta t = 0.005$ ms) at the three specified spatial resolutions was considered. The first observation is that for each configuration, the number of updates per second is constant, independent of the number of nodes. Since the total number of timesteps is constant for all resolutions, this reflects the linear scaling of the algorithm with the number of nodes. The second observation is the speedup obtained with different forms of parallelization. The OpenMP version with 4 executing threads was 3 times faster than the serial version. The GPU version was almost 12 times faster than the OpenMP version, resulting in a total speedup of 35 times over the single processor version.

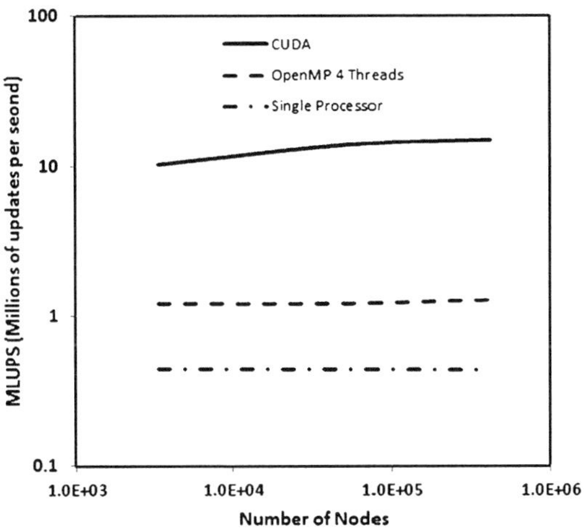

Figure 2.15. Computational performance of the LBM-EP algorithm on different architectures: single processor, multicore processing on the CPU and on the graphical processing unit.

2.2.2 Efficient modeling of the electrical conduction system

As discussed in section 1.2.2, the electrical conduction system is a complex anatomical structure, whose geometrical and functional properties determine the pattern of ventricular excitation and contraction. To properly reproduce the sequence of cardiac activation, a ventricular electrophysiology model must be able to accurately capture the location and thickness of the Purkinje system, including patient-specific information whenever available. The effect of high-speed bundles can be modeled as a localized increase of the electrical conductivity of the myocardial tissue. Assuming that the left and right bundle branches, as well as the Purkinje fibers, are densely diffused in the subendocardium, as reported for instance in [58], a rule can be defined to classify the myocardial tissue as part of the high-speed bundles, based on its distance from the endocardium. Numerical methods based on Cartesian grids pose a challenge to this approach, since the rasterization process limits the spatial accuracy of all space-dependent quantities to the resolution of the grid. After the rasterization, grid nodes are classified as either part of the high-speed conducting tissue or part of the normal tissue. Therefore, the sub-endocardial layer is approximated with an error that depends on the grid resolution. This can be a significant limitation when describing the Purkinje system, that extends subendocardially in a layer whose thickness may be of the same order of magnitude of the grid spacing (up to one third or one half of the ventricular thickness). Specialized methods can be designed, able to provide sub-grid spatial accuracy, thus reproducing the same effect of high-speed conducting tissue independently of the resolution of the Cartesian grid.

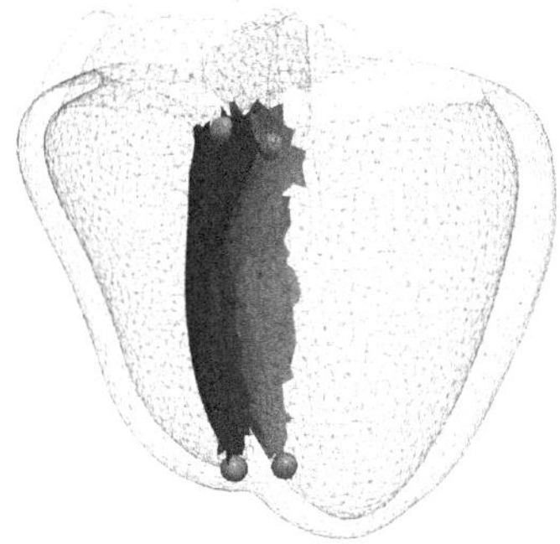

Figure 2.16. Illustration of the spontaneous activation points.

As a preliminary step, anatomical landmarks are automatically identified on the patient-specific heart as anchors of the cardiac activation model. In particular, the septal portion of the right and left endocardium is considered. Spontaneus activation points located on the ventricular septum are computed from geometric landmarks, as shown in Fig. 2.16. A level set function ϕ can be used to define the surface of interest and the distance of a point from the surface. For the sake of clarity, in the following we consider the Euclidean distance to the closest surface point. Furthermore, the surface is assumed to be equipped with a field of normal unit vectors, so that one can identify internal and external points easily. Internal points are those with negative distance from the surface, and external points are characterized by a positive distance from the surface. When referring to the endocardial surface, internal points are in the blood pool while external points are in the myocardium.

Myocardial tissue is classified as high-speed conducting if the distance from the endocardium is positive and smaller than a user-defined threshold h. For every point $\mathbf{x}$ in the myocardium,

selection criterion:

$$\begin{cases} 0 \leq \phi(\mathbf{x}) \leq h & \rightarrow \text{High-speed conducting tissue} \\ \phi(\mathbf{x}) > h & \rightarrow \text{Normal tissue.} \end{cases} \tag{2.6}$$

Tissue conductivity σ is modeled as a piecewise constant scalar field over the Cartesian grid. Each grid point is at the center of a voxel – a very small volume of tissue. The conductivity value assigned to each grid point ranges from normal to high based on the volume fraction ψ of tissue in the voxel whose distance from the endocardium is smaller than the threshold h. For every lattice node $\mathbf{x}$ we have

$$\sigma(\mathbf{x}) = \psi\,\sigma_{\text{purkinje}} + (1-\psi)\,\sigma_{\text{normal}}.$$

The volume fraction ψ is computed by a two-step algorithm. In the first phase, a list of candidates, i.e., grid points that may have a partial volume of high-speed conducting tissue, is built. Nodes that are closer to the endocardium than the target distance h are included in the list, as well as nodes that are further away but may have at least part of the boundary within the high-speed conducting tissue. In other words, all nodes whose distance from the endocardium is less than an extended threshold h_{ext} are selected, corresponding to the thickness of the layer of high-speed conducting tissue plus the maximum distance between the barycenter of the voxel and its boundary.

$$h_{\text{ext}} = h + \frac{\sqrt{3}}{2}\Delta x$$

Δx being the spacing of the lattice. Given ϕ, level-set representation of the endocardial surface, its discretization $\phi_{\Delta x}$ over the Cartesian grid is computed. All lattice nodes $\widehat{\mathbf{x}}$ such that $\phi_{\Delta x}(\widehat{\mathbf{x}}) < h_{\text{ext}}$ are selected.

In the second phase, for each candidate we consider the voxel $\widehat{\mathbf{v}}$ centered in $\widehat{\mathbf{x}}$ and define a sub-grid of nodes $\boldsymbol{\xi} \in \widehat{\mathbf{v}}$ with uniform spacing $\Delta\xi < \Delta x$. We compute $\widehat{\phi_{\Delta\xi}} = \phi|_{\widehat{\mathbf{v}}}$, the discretization of the level set function on the sub-grid and use it to classify the nodes of the sub-grid as normal or high-speed conducting tissue, based on their distance from the endocardium:

$$\forall\, \boldsymbol{\xi} \in \widehat{\mathbf{v}}: \begin{cases} 0 \le \widehat{\phi_{\ell}}(\boldsymbol{\xi}) \le h \rightarrow \text{High-speed conducting tissue} \\ \widehat{\phi_{\Delta\xi}}(\boldsymbol{\xi}) > h \rightarrow \text{Normal tissue.} \end{cases}$$

Finally, the partial volume ψ is obtained as the number of sub-grid nodes belonging to the high-speed tissue over the total number of sub-grid nodes. Algorithm 2 summarizes the main steps of the method.

A key strength of this approach is that it allows an accurate evaluation of the partial volume of tissue within a given distance from the surface, especially when the spacing of the original lattice

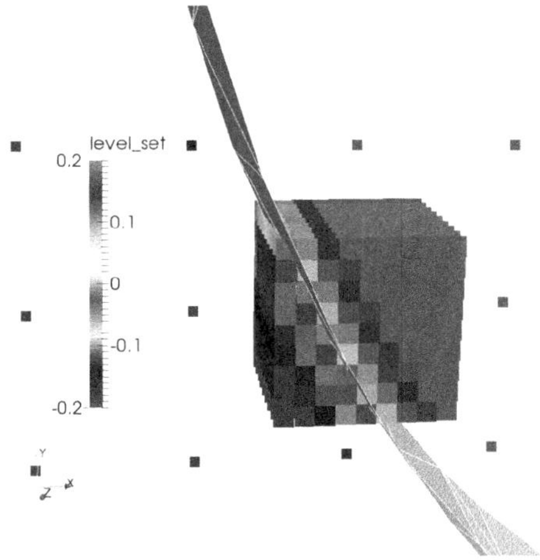

Figure 2.17. Graphical representation of the modeling approach for high-speed conducting tissue. The lattice nodes of the Cartesian grid are shown in the background, colored by the local value of the level set. The triangulated surface represents the endocardium. For one of the lattice nodes, the sub-grid defined in the voxel is visualized, each point in the sub-grid being colored by the value of the level set. In this example we considered a threshold $h = 0.1$ mm. For visualization purposes, the color bar has been scaled to the interval $[-0.2, 0.2]$ mm.

is of the same order of magnitude of the threshold h, or larger. As shown in Fig. 2.17, if the conductivity in the voxel would have been assigned based on its distance from the surface (as evaluated on the original lattice), the entire volume of tissue would have been classified as high-speed conducting. As a further straightforward extension of the method, the thickness h can be space-dependent to take into account variations in the spatial distribution of high-speed conducting tissue.

Another important feature of this method is that the surface represented by the level set ϕ is generic. Any patient-specific geometric information about the Purkinje system, if available, can be readily included in the model. For instance, if the boundaries of high-speed conducting tissue can be segmented from patient-specific medical images, in terms of a boundary surface equipped with normal unit vectors, then the method can be applied by modifying the selection criterion (Eq. (2.6)), setting $h = 0$ and defining internal points as having negative distance from the surface:

$$\text{selection criterion:} \begin{cases} \phi(\mathbf{x}) \leq 0 & \rightarrow \text{High-speed conducting tissue} \\ \phi(\mathbf{x}) > 0 & \rightarrow \text{Normal tissue.} \end{cases}$$

Algorithm 2 Modeling high-speed conducting tissue with a space-dependent conductivity field σ.

Require: Volume V discretized in a Cartesian grid with N_h points $\mathbf{x}_i$, level-set representation ϕ of septal endocardium, user-defined threshold h, grid spacings Δx and $\Delta \xi$, conductivity for normal tissue σ_{normal} and for high-speed conducting system σ_{purkinje}.

$\phi_h = \phi|_V$
candidates := {}
$h_{\text{ext}} := h + \frac{\sqrt{3}}{2}\Delta x$
for $i = 1 \to N_h$ **do**
    if $0 \leq \phi_h(\mathbf{x}_i) \leq h_{\text{ext}}$ **then**
        candidates = { candidates, i }
    end if
end for
for $j \in$ candidates **do**
    count $= 0$;
    compute $\widehat{\phi}_{\Delta\xi} = \phi|_{\widehat{\mathbf{v}}_j}$ subdividing voxel $\widehat{\mathbf{v}}_j$ in $N_{\Delta\xi}$ lattice points $\boldsymbol{\xi}_k$
    for $k = 1 \to N_{\Delta\xi}$ **do**
        if $0 \leq \phi_{\Delta\xi}(\boldsymbol{\xi}_k) \leq t$ **then**
            ++count;
        end if
    end for
    $\psi_j =$ count $/N_{\Delta\xi}$
    $\sigma(\mathbf{x}_j) = \psi\, \sigma_{\text{purkinje}} + (1 - \psi)\, \sigma_{\text{normal}}$
end for
return Σ

2.2.3 Graph-EP: fast computation of tissue activation time

When the focus is on the very fast computation of the pattern of electrical activation, and it is not required to model the details of cellular electrophysiology, Eikonal models offer a convenient alternative to monodomain or bidomain models. The Eikonal equation describes the propagation of a wave front in a domain Ω, given an initial configuration on a subdomain $\Gamma \subset \Omega$:

$$\begin{cases} \sqrt{\nabla a^T \mathbf{R} \nabla a} = 1 & \text{in } \Omega \\ a = a_0 & \text{on } \Gamma \end{cases} \tag{2.7}$$

where the unknown a is the activation time, Ω represents the myocardium, Γ represent the regions of earliest activation in the myocardium and R is a symmetric conductivity tensor:

$$R = \sigma^2 \left[(1-\rho)\mathbf{f}\mathbf{f}^T + \rho \mathrm{I} \right] \tag{2.8}$$

σ is the local conduction velocity (along the fiber direction), $\rho \in [0, 1]$ is the anisotropy ratio, I the identity matrix and **f** the fiber direction.

A very efficient numerical solution of the Eikonal equation is based on a variant of Dijkstra's shortest path algorithm [82]. The input of the method is the tetrahedral mesh of the patient's heart, which is tagged to identify where the EP wave starts (left and right ventricular septum, see also Fig. 2.16). As a first step, the nodes from where the electrical wave starts are added to a priority queue, with a value equal to their activation time. For instance, the LV septum nodes are added with their prescribed activation time a_{LV}, and the RV septum nodes with their prescribed activation time a_{RV}. Any additional activation point (e.g. due to device pacing) can be added to the priority queue based on its own activation time. The first node of the queue is then selected and all its neighbors are processed.

Let n_i be the node which is currently processed, and n_j one of its neighbors. A tentative activation time a'_j is computed as $a'_j = a_i + t_{ij}$, where t_{ij} is the traveling time from node i to j given by the formula $t_{ij} = c_{ij}/\sigma_{ij}^{tissue}$, c_{ij} is the edge cost and σ_{ij}^{tissue} is the conduction velocity of the tissue crossed by the edge $n_i n_j$, calculated using weighted average. If a'_j is smaller than the current activation time estimate a_j, then a_j is updated ($a_j = a'_j$), and the node n_j is placed in the queue for further processing. The process is then iterated until no more nodes need to be processed, meaning that the heart graph is fully processed. Tissue anisotropy is modeled by modifying the edge cost c_{ij} to take into account fiber orientation as follows:

$$c_{ij} = \frac{\sqrt{\mathbf{n_i n_j}^T \left[(1-\rho)\mathbf{f}_{ij}\mathbf{f}_{ij}^T + \rho \mathrm{I} \right] \mathbf{n_i n_j}}}{\delta_{ij}} \tag{2.9}$$

where $\mathbf{n_i n_j}$ is the edge vector and δ_{ij} is the edge length. In our case, since the fibers are defined node wise, we define $\mathbf{f}_{ij} = (\mathbf{f}_i + \mathbf{f}_j)/2$ as the local edge-based approximation of the fiber vector, used to compute R. The Purkinje network is handled in the model by edge-based linear interpolation of the conduction velocities. Finally, the transmembrane potentials $v(t)$ are approximated by assigning at a given time t a value of -70 mV to the nodes that are still not activated, $+30$ mV otherwise.

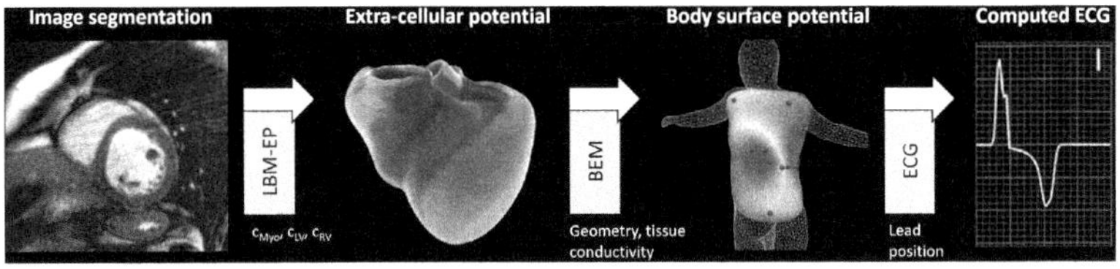

Figure 2.18. Overview of the workflow for computational modeling of patient-specific ECG.

2.2.4 Body surface potential modeling

A two-step procedure is used to compute electrocardiograms based on the computed transmembrane potentials $v(t)$. First, extra-cellular potentials are estimated, which can be computed using LBM-EP or Graph-EP. In a second step, we employ a Boundary Element Method (BEM) to project the extra-cellular potentials onto the torso mesh and derive the ECG leads. Fig. 2.18 illustrates the complete workflow of the forward model from images to ECG signals.

Extracellular potentials computation

Defining the trans-membrane potentials as the difference between intra- and extracellular potentials ($v = \phi_i - \phi_e$), Chhay et al. [99] proposed the following formulation of the bidomain problem to recover $\phi_e(t)$ from $v(t)$:

$$\alpha[\gamma \partial_t v + J_{total}(v, w)] = \nabla \cdot (\mathrm{R}_i \nabla (v + \phi_e)), \tag{2.10}$$

$$\nabla \cdot ((\mathrm{R}_i + \mathrm{R}_e)\nabla \phi_e) + \nabla \cdot (\mathrm{R}_i \nabla v) = 0. \tag{2.11}$$

The parameters α and γ are, respectively, the ratio of surface of membrane per unit volume and the membrane capacitance per unit area. R_i and R_e are intra- and extracellular conductivity tensors, while J_{total} denotes the total ionic current. Finally, the influence of the selected cell model is modeled via the state variable w, which would correspond to the gating variable h if the considered cellular model is the M-S model. Assuming a constant diffusion anisotropy ratio λ at any position $\mathbf{x}$, $\lambda = \mathrm{R}_i(\mathbf{x})/\mathrm{R}_e(\mathbf{x})$, one can define a lumped diffusion tensor $\mathrm{R}(\mathbf{x}) = \frac{1}{1+\lambda}\mathrm{R}_i(\mathbf{x})$, and rewrite the parabolic system more succinctly as:

$$\alpha[\gamma \partial_t v + J_{total}(v)] = \nabla \cdot (\mathrm{R}\nabla(v)), \tag{2.12}$$

with the boundary condition $\mathrm{R}\nabla v \cdot \mathbf{n} = 0$ on the boundary of the computational domain ($\mathbf{n}$ is the epicardial surface normal). Under this assumption, the extra-cellular potential can be computed using the following closed-form expression (Ω defines the computational domain; $|\Omega|$ is its volume) [99]:

$$\phi_e(\mathbf{x}, t) = \frac{\lambda}{1+\lambda} \frac{1}{|\Omega|} \int_{\Omega} [v(y,t) - v(\mathbf{x}, t)] dy. \tag{2.13}$$

In case the extra-cellular potential is computed on a Cartesian grid, as done with the LBM method, it can be easily mapped back onto the original epicardial mesh using tri-linear interpolation.

Boundary element model of torso potentials

The boundary element method (BEM) is commonly used to map extracellular potentials onto a torso mesh by solving the Laplace equation $\nabla \cdot \mathrm{R}_T \nabla \phi = 0$ with a Neumann boundary condition $\mathrm{R}_T \nabla \phi \cdot \mathbf{n} = \mathbf{0}$ on the torso mesh S_B, and the Dirichlet boundary condition $\phi = \phi_e$ on the epicardium S_H. The parameter R_T denotes the tissue-dependent conductivity tensor.

Green's second identity writes $\int_S (A\nabla B - B\nabla A) \cdot \mathbf{n} dS = \int_V (A\Delta B - B\Delta A) \cdot dV$ for a volume V, its boundary surface S and normal vector $\mathbf{n}$, where A and B in this equation are scalar functions of position. By defining A as the product of the electric potential and the isotropic conductivity, and B as the term $1/\mathbf{r}$, where $\mathbf{r}$ is the vector from a particular integration point to the position under investigation, Green's second identity can be used to analyze voltages on the surface of a conducting volume [222].

A personalized model that includes variable thoracic cavity conductivity would be the ideal setup for accurate ECG simulations. However, in a first approximation, one may assume an isotropic conductivity between epicardium and body surface. Therefore, the potential at any point x of the thoracic domain can be formulated as follows:

$$\phi(x) = \frac{1}{4\pi} \int_{S_B} \phi_b \frac{\mathbf{r} \cdot \mathbf{n}}{\|\mathbf{r}\|^3} dS_B - \frac{1}{4\pi} \int_{S_H} \left[\phi_e \frac{\mathbf{r} \cdot \mathbf{n}}{\|\mathbf{r}\|^3} + \frac{\nabla \phi_e \cdot \mathbf{n}}{\|\mathbf{r}\|} \right] dS_H. \tag{2.14}$$

Here ϕ_e denotes the previously computed extracellular potentials at the epicardium, while ϕ_b are the unknown body surface potentials on the torso. If the surfaces S_B (torso) and S_H (epicardium) are discretized into triangular meshes, the problem can be formulated as the solution of a linear system of equations [222]:

$$\mathrm{P}_{BB}\phi_b + \mathrm{P}_{BH}\phi_e + \mathrm{G}_{BH}\Gamma_H = 0,$$
$$\mathrm{P}_{HB}\phi_b + \mathrm{P}_{HH}\phi_e + \mathrm{G}_{HH}\Gamma_H = 0.$$

The matrix Γ_H collects the gradients $\nabla\phi_e$ and does not require extra computation. The matrices P and G contain coefficients depending uniquely on the geometry of the epicardium and the body torso meshes, and can therefore be precomputed. Their rows correspond to different locations of the observation point on the surface indicated by the first subscript (B: body surface, H: heart). Similarly, their columns correspond to different locations on the surface of integration, indicated by the second subscript.

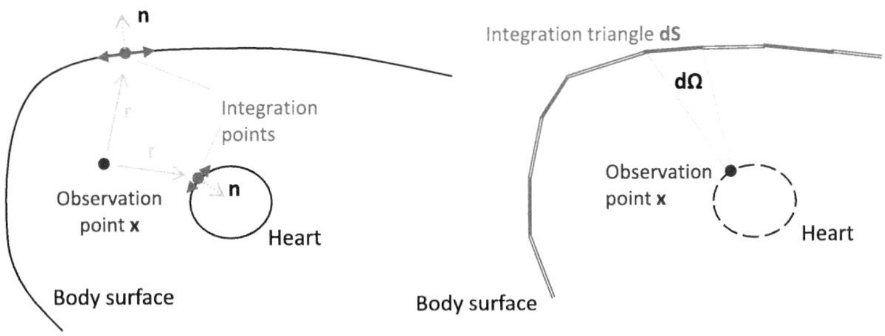

Figure 2.19. Left Panel: general principle of geometry matrix definition using the example of P_{HB} (observation points on the heart and torso as integration surface), see text for details. Right panel: actual implementation principle of P_{HB} calculation using triangular mesh representation of body surface. Different triangles on the torso are represented using different shades of red (mid gray in print version).

Following the formulation of [222], the matrices are constructed as follows:

- The P matrices contain coefficients regarding the potentials ϕ_b and ϕ_e. For a given triangle k of the integration surface S, and a particular observation point i on the observation surface, the coefficient is expressed as the solid angle subtended by triangle k onto point i: $P_k^i = \int_{S^k} d\Omega_S$. These P_k^i coefficients need to be distributed onto the three vertices of triangle k, in order to compute element P_{ij} of the desired matrix, as illustrated below.
- The G matrices contain coefficients regarding the normal components of the gradients in Γ_H. Using the same notation as before, we can define the coefficients $G_k^i = \int_{S^k} \frac{dS}{r^i}$, where r^i is the distance from the observer's location to the respective point on the integration surface. The integral is computed using the assumption that the triangle k can be approximated by a circle sector that is perpendicular to the direction of r^i: $\int_S \frac{dS}{r} = \sqrt{\theta^2 d^2 + 2A\theta} - \theta d$, where A is the area of the triangle, d the distance to its closest vertex with respect to the observation point, and θ the angle between the two triangle edges sharing the closest vertex.

In both cases, special care has to be taken if integration and observation surface are identical (i.e., matrices P_{BB}, P_{HH} and G_{HH}), in which case partitioning into a close region (triangles around the integration point) and a distant region (all other triangles, Fig. 2.19) is reasonable, as it simplifies computation significantly.

Finally, we define the transformation matrix

$$Z_{BH} = [P_{BB} - G_{BH}G_{HH}^{-1}P_{HB}]^{-1}[G_{BH}G_{HH}^{-1} - P_{BH}]$$

which allows us to compute body surface potentials directly from the extracellular potentials of the heart by means of a single matrix multiplication: $\phi_b = Z_{BH}\phi_e$.

2.2.4.1 ECG calculation

Based on the body surface potentials, which are computed for each vertex at the torso mesh, we compute the standard Einthoven and Goldberger limb leads (I, II, III, a_{VR}, a_{VL}, a_{VF}) as well as the Wilson precordial leads (V_1–V_6). To avoid interpolation on the torso mesh, the electrode positions can be chosen to coincide with vertex positions. The following ECG features are automatically detected as in [223]:

QRS duration $QRSd$**:** For numerical stability, the QRS complex is detected using the depolarization times computed by the tissue level electrophysiology model. Assuming one full heart cycle is computed, $QRSd = max_{\mathbf{x}} a(\mathbf{x}) - min_{\mathbf{x}} a(\mathbf{x})$. The depolarization (or activation) times a are obtained as the points in time when the potential first exceeds the change-over voltage:

$$a(\mathbf{x}) = \text{argmin}_t \{v(\mathbf{x}, t) \geq v_{gate}\}$$

Electrical Axis EA**:** For the limb leads I and II, the peak amplitudes h_I and h_{II} are computed by summing up the amplitudes of R and S waves in the respective leads. The electrical axis is then calculated using the formula $EA = \arctan \frac{2h_{II} - h_I}{\sqrt{3}h_I}$.

2.3 Biomechanics modeling

The next modeling component is cardiac biomechanics. The goal is to efficiently solve Newton's law of motion for the myocardium (Eq. (2.15)), given patient-specific anatomy and electrophysiology:

$$M\ddot{\mathbf{u}} + D\dot{\mathbf{u}} + K(\mathbf{u})\mathbf{u} = \mathbf{f}_a + \mathbf{f}_p + \mathbf{f}_b \tag{2.15}$$

with $\mathbf{u}$ the vector of nodal displacements, $\mathbf{f}_a$ the active force of the depolarized cells, $\mathbf{f}_{bp}$ the endocardial pressure, $\mathbf{f}_b$ the external forces and $\mathrm{K}(\mathbf{u})$ the stiffness matrix, non-linearly dependent on the deformation. D represents a Rayleigh-type damping, i.e. $\mathrm{D} = \lambda_M \mathrm{M} + \lambda_K \mathrm{K}$, with λ_M and λ_K user-defined scalar coefficients.

The dynamics equaton (2.15) is traditionally solved by using the finite element method (FEM) with a hexahedral or tetrahedral spatial discretization of the myocardial domain, although finite difference [130] and mesh-less [224,225] approaches have also been explored. Using the FEM, the equation is solved on the mesh nodes, with the nodal forces calculated by accumulating the contributions of all elements sharing common nodes. A particularly efficient algorithm, using the FEM and explicit central difference temporal discretization of Eq. (2.15) (see e.g. [104], p. 803), is the Total Lagrangian Explicit Dynamics (TLED) algorithm, originally proposed by Miller et al. [106]. Its main idea is to refer all mathematical quantities of interest to the original undeformed configuration of the system. Such a formulation allows to pre-compute several quantities, thus reducing the amount of mathematical operations to be done at each time step. Furthermore, TLED uses low order polynomial approximation for the unknowns of the equation and a lumped mass approximation [104] (such that M is a diagonal matrix). This, combined with the explicit time integration scheme, results in computationally inexpensive time stepping. At each time step no matrix inversion is required (as would be the case for implicit or semi-implicit time advancing algorithms) and operations are node-wise, hence massively parallelizable on GPU for instance. For these reasons, TLED is particularly suitable for real-time computation of soft tissue deformations like brain, liver, kidney [106]. Mathematical formulations for the cardiac use case are provided below, following the approach presented in [226].

Let ϕ be the differentiable material deformation function, $\mathrm{F} = \nabla\phi$ the deformation gradient, $J = \det \mathrm{F}$ the Jacobian determinant, and $\mathrm{C} = \mathrm{F}^T\mathrm{F}$ the right Cauchy–Green tensor. Myocardium biomechanics is modeled using a Hill–Maxwell framework (section 1.3), which decomposes the total Cauchy stress T into the sum of its active and passive parts, $\mathrm{T} = \mathrm{T}_a - \mathrm{T}_p$. The negative sign associated to the passive stress appears as the stiffness matrix $\mathrm{K}(\mathbf{u})$ is in the left-hand side of Eq. (2.15), whereas the active stress is in the right-hand side, along with all other forces. The combined force for each element is then obtained by integrating the contribution of the second Piola-Kirchhoff (SPK) stress S

$$\mathbf{f} = \mathbf{f}_a - \mathbf{f}_p = \int_{V_0} \mathrm{Z}^T \mathrm{S}\, dV_0 \tag{2.16}$$

where Z is the strain-displacement matrix [104], V_0 is the volume of the element at time $t = 0$, and S is the SPK tensor expressed in terms of the Cauchy stress tensor as

$$\mathrm{S} = J\mathrm{F}^{-1}\left(\mathrm{T}_a - \mathrm{T}_p\right)\mathrm{F}^{-T}. \tag{2.17}$$

The next sections describe how each force is computed. In [226] the authors provide a more detailed description of the algorithm, including the computation of the strain-displacement operator Z from the finite element discretization.

2.3.1 Passive stress component

As introduced in section 1.3.1, the simplified, transverse isotropic Holzapfel–Ogden model energy function writes:

$$\psi = \frac{a}{2b} exp[b(I_1 - 3)] + \frac{a_f}{2b_f}\left\{exp[b_f(I_{4f} - 1)^2] - 1\right\} + d_1(J - 1)^2. \tag{2.18}$$

The term $d_1(J - 1)^2$ models incompressibility. The parameter d_1 is therefore equivalent to a bulk modulus. Deriving the invariants with respect to C gives $\partial I_1/\partial \mathrm{C} = \mathrm{I}$ and $\partial I_{4f}/\partial \mathrm{C} = \mathbf{ff}^T$. As a result, the Cauchy stress writes as follows, where we introduced a stiffness parameter β to simplify model personalization (see section 2.5.3):

$$\begin{aligned} \mathrm{T}_p = \beta &\left(\frac{a}{2}\exp[b(I_1 - 3)]\,\mathrm{I} + a_f \exp[b_f(I_{4f} - 1)^2]\,\mathbf{ff}^T\right) \\ &+ d_1(J - 1)J\mathrm{C}^{-1}. \end{aligned} \tag{2.19}$$

2.3.2 Active stress component

In this implementation, a simplified, phenomenological model is preferred for its computational efficiency while being globally fidel to the underlying mechanisms of myocyte contraction [45]. The active stress T_a is controlled by a switch function $u(t)$, related to the activation time, here denoted t_d, and the repolarization time, denoted t_r, computed by the cardiac electrophysiology model (section 2.2, Fig. 2.20). When the cell is depolarized ($t_d \le t < t_r$), $u(t)$ is constant and is equal to the contraction rate $+k_{ATP}$. This variable relates to the rate of ATP consumption by the sarcomere that fuels the CICR mechanism of the myofilaments (see section 1.3.2), as well as the number of cross-bridges recruited to perform the contraction. When the cell is repolarized ($t_r \le t < t_d + CL$, CL is the cycle length of the myocyte), $u(t)$ is equal to the relaxation rate $-k_{RS}$, which relates to the rate of unbinding of the cross-bridge, and hence the decrease in contraction force.

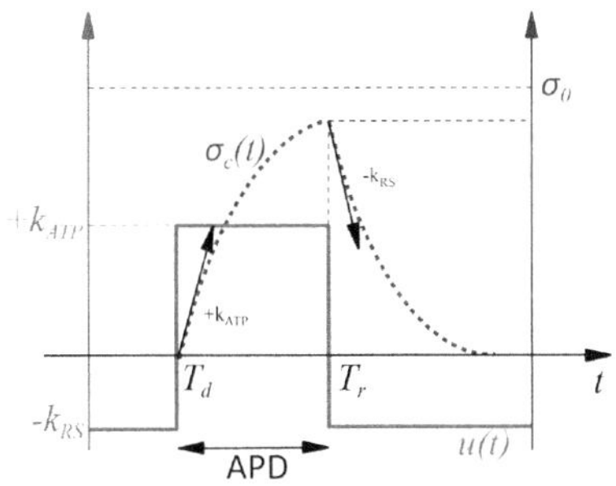

Figure 2.20. Variation of the active contraction stress $\tau_c(t)$ (in blue (dark gray in print version)) with respect to the electrical command function $u(t)$ (in red (mid gray in print version)) controlled by the cardiac electrophysiology model.

Let τ_0 be the maximal stress one cell can generate if all the cross-bridges are recruited, and $|u(t)|_+$ the positive part of the function $u(t)$. The change in cell stress over time, denoted $\tau_c(t)$, is modeled by the ODE:

$$\frac{d\tau_c(t)}{dt} + |u(t)|_+ \tau_c(t) = |u(t)|_+ \tau_0. \tag{2.20}$$

An analogous equation, but without the constant term on the right hand side, holds with the negative part of the function $u(t)$. These equations can be solved analytically, giving the following closed form solutions:

$$\begin{cases} \text{if } t_d \leq t < t_r : \tau_c(t) = \tau_0 \left[1 - e^{+k_{ATP}(t_d - t)}\right], \\ \text{if } t_r \leq t < t_d + CL : \tau_c(t) = \tau_c(t_r) e^{-k_{RS}(t_r - t)}. \end{cases} \tag{2.21}$$

The contraction is mostly performed along the fiber direction **f**. As a result, the active stress writes $\mathrm{T}_a = \tau_c \mathbf{f}\mathbf{f}^T$, and is integrated into the SPK tensor following Eq. (2.17).

In summary, the model is controlled by three families of free parameters that can be defined either globally (e.g. one value per ventricle), or at each node of the mesh:

- τ_0: maximal strength of the active contraction
- k_{ATP}: rate of contraction that controls the speed at which the muscle contracts
- k_{RS}: rate of relaxation that controls the speed at which the muscle relaxes.

2.3.3 Myocardial boundary conditions

Endocardial pressure

The intra-ventricular blood pressure is computed using a lumped parameter model of blood flow coupled with the ventricles through the pressure unknown $p(t)$. Fig. 2.21 shows the pressure-flow model used in this chapter, with coupled valve, artery, ventricle, and atrium modules. In brief, there is a two-way interaction between the remote (arterial and atrial) systems, which receive flow (positive or negative, depending on the time in the cardiac cycle) from the valve modules, and the ventricles, which receive pressure values used as boundary conditions from the arterial and atrial systems.

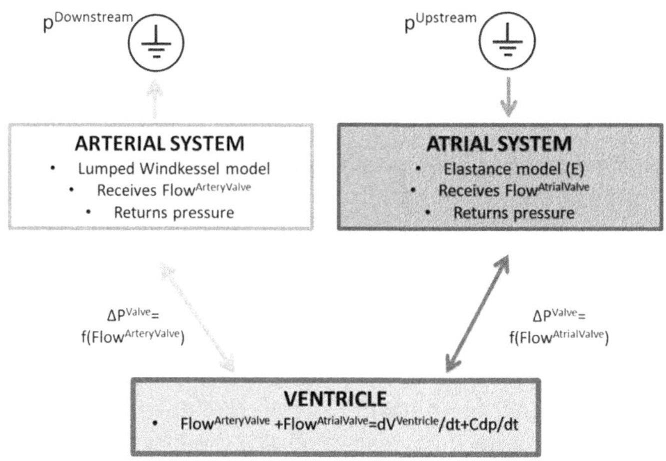

Figure 2.21. Cardiac pressure-flow system with pressure driven valves that modulate the interactions between the ventricles, arteries and atria. The remote pressures can be set independently to physiological values or can be connected as part of a whole body circulation system.

The mass balance of the ventricular volume $V(t)$ can be specified in such a way that the isovolumic stages of the cardiac cycle, where there is no flow through valves and therefore the pressure variable $p(t)$ is an unknown of the coupled system, are handled by the same equation:

$$\frac{dV}{dt} = \phi^{\text{arterial}} + \phi^{\text{atrial}} - \mu \frac{dp}{dt} \tag{2.22}$$

ϕ^{arterial} and ϕ^{atrial} are the bulk flow rates across the valves, expressed from the ventricle point of view. They are treated as independent variables and handled by the valve modules (Fig. 2.21), which receive pressures as inputs from both sides (ventricle/artery or ventricle/atrium) and output the flow across the valves, as described in more detail in section 2.4. The valves are pressure-driven, meaning that they open and close according to the pressure gradient through them. The last term of Eq. (2.22) is an "artificial compressibility" term [227] controlled by the capacitance parameter μ. That term transforms the minute temporal changes in the ventricular volume $V(t)$, computed by the biomechanical solver during isovolumic stages, into temporal pressure gradients.

The final balance of these flows (Eq. (2.22)) with the ventricular flow rate dV/dt, computed directly from the finite element mesh, allows direct computation of the ventricular pressure by integrating dp/dt. The boundary face traction is finally updated using the computed pressure $\mathbf{f}_p = -p\mathbf{n}$, where $\mathbf{n}$ is the endocardial surface normal.

Attachment to atria and arteries

The effect of arteries and atria on the ventricular motion is simulated by connecting the vertices of the valve plane to springs whose stiffness is k_{base}. The fixed extremity of the springs corresponds to the rest position of the nodes, taken at mid diastasis, when the heart is at rest. The spring stiffness k_{base} is anisotropic to allow free in-plane motion $(\mathbf{e}_r, \mathbf{e}_c)$ while minimizing the longitudinal motion of the base along the long axis $\mathbf{e}_l$ of the heart. Under these definitions, the basis stiffness force writes:

$$\mathbf{f}_{\text{base}} = \mathrm{M}^{-1} \begin{pmatrix} k_{\text{base},l} & 0 & 0 \\ 0 & k_{\text{base},r} & 0 \\ 0 & 0 & k_{\text{base},c} \end{pmatrix} \mathrm{M}(\mathbf{x} - \mathbf{x}_0) \tag{2.23}$$

where M is the transformation matrix going from the global coordinate system to the coordinate system defined by the LV long axis and the short axis plane, as illustrated in Fig. 2.22.

Modeling the effect of the pericardium bag

In [6,135], a contact-based model of the pericardium has been proposed to mimic the effects of the neighboring organs and of the pericardium bag on the cardiac motion. The idea consists in reducing the motion of the epicardial nodes in the radial direction towards the outside of the pericardium, while allowing friction-free sliding. For this, the pericardial domain is first estimated as a signed distance map $\Pi(\mathbf{x})$ from the detected cardiac epicardium at end-diastole (Fig. 2.23, left panel). The interior and exterior of the pericardium bag are negative and positive regions respectively

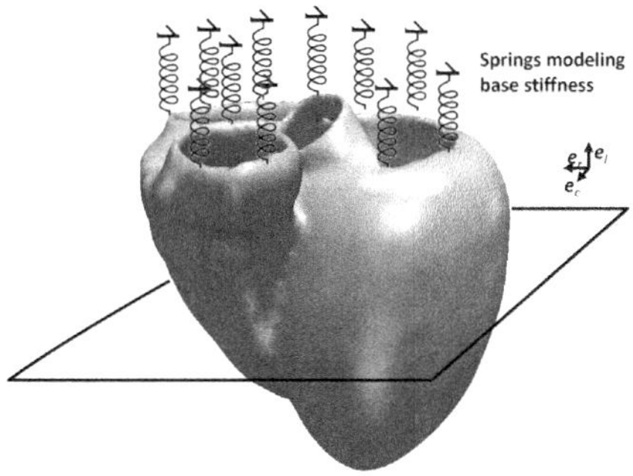

Figure 2.22. The effect of arteries and atria on the ventricles is modeled by attaching springs to the valve plane.

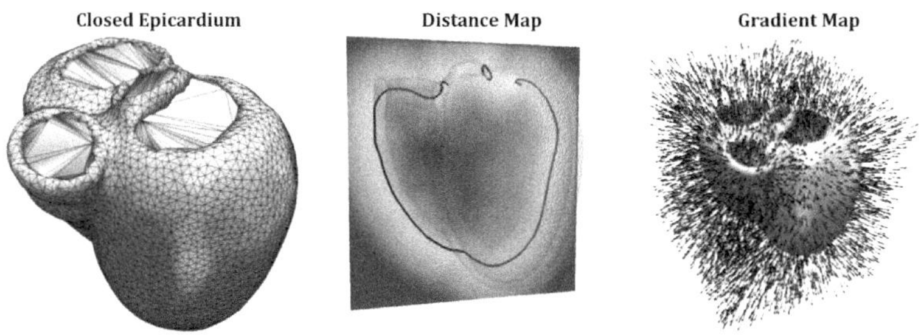

Figure 2.23. The pericardium region is defined by the epicardium at end diastole. *Left panel*: the valves are closed automatically to define an inner and outer region. A distance map is built (*middle panel*), to identify regions in which the epicardium is allowed to freely slide along the pericardial bag (authorized region in white, epicardium in black). As soon as an epicardial node goes outside the authorized region, a force along the gradient map (*right panel*) is applied to bring that node back.

(Fig. 2.23, mid panel). The pericardial contact forces are then defined as:

$$\begin{cases} \mathbf{f}_{\text{peri}}(\mathbf{x}) & = -k_{out} \frac{\Pi(\mathbf{x})^2}{\Pi(\mathbf{x})^2+m^2} \nabla(\Pi(\mathbf{x})), \text{ if } \Pi(\mathbf{x}) > d_{out} \\ \mathbf{f}_{\text{peri}}(\mathbf{x}) & = k_{in} \frac{\Pi(\mathbf{x})^2}{\Pi(\mathbf{x})^2+m^2} \nabla(\Pi(\mathbf{x})), \text{ if } \Pi(\mathbf{x}) < d_{in} \\ \mathbf{f}_{\text{peri}}(\mathbf{x}) & = 0, \text{ otherwise} \end{cases} \tag{2.24}$$

where, d_{in} and d_{out} are thickness parameters that control the extent of the constraining region. The stiffness of the contact is controlled by the parameters k_{in} and k_{out}. The parameter m finally controls how fast the maximum strength of the contact force is reached. With this definition, the epicardial nodes can slide along the pericardium cavity, but their radial motion is limited to emulate neighboring organs and the stiff pericardial sac. Finally, the total boundary condition force is given by $\mathbf{f}_b = \mathbf{f}_{\text{base}} + \mathbf{f}_{\text{peri}}$.

2.3.4 Putting it all together: a fast computational framework for cardiac biomechanics

Combining all the forces derived in the previous sections and distributing them to the nodes of the mesh leads to the final TLED formulation. Time integration is done using the central difference scheme:

$$\mathbf{u}(t+\delta t) = \mathbf{u}(t) + \delta t \dot{\mathbf{u}}(t) + \frac{1}{2}\delta t^2 \ddot{\mathbf{u}}(t). \tag{2.25}$$

Let us assume that the displacements $\mathbf{u}(t)$ and $\mathbf{u}(t-\delta t)$ are known, and that the total nodal forces $\mathbf{f}(t)$ have been computed (including internal and external forces). The central difference method then yields a node-wise equation for computing displacements at the next time step:

$$\mathbf{u}_i(t+\delta t) = A_i \mathbf{f}_i(t) + B_i \mathbf{u}_i(t) + C_i \mathbf{u}_i(t-\delta t), \tag{2.26}$$

$$A_i = \frac{1}{\frac{\mathrm{D}_{ii}}{2\delta t} + \frac{\mathrm{M}_{ii}}{\delta t^2}}, \quad B_i = \frac{2\mathrm{M}_{ii}}{\delta t^2} A_i, \quad C_i = \frac{\mathrm{D}_{ii}}{2\delta t} A_i - \frac{B_i}{2}. \tag{2.27}$$

As any explicit time integration, δt must be small enough to guarantee stability and accuracy. In particular, δt must be smaller than the critical limit $\delta t_{cr} = L_e / c$, where L_e is the smallest characteristic element length in the assembly and c is the dilatational wave speed of the material.

Description of the TLED finite elements algorithm

One of the advantages of TLED algorithm is that all the variables are referred to the initial configuration, which allows very efficient updates during the main iteration loop. The algorithm starts with a pre-computation stage to initialize all variables (Algorithm 3). Before starting the main iteration loop, the solver and constraints are initialized (Algorithm 4), followed by the main iteration, which consists in first calculating the net nodal force (Algorithm 5) and then updating the displacement (Algorithm 6).

2.3.5 Evaluation of the TLED algorithm

Validation against analytical solution

As shown in [120], simple shear in different planes can be used to verify the implementation of any constitutive law. Indeed, this

Algorithm 3 TLED algorithm: pre-computation of constant variables.

Require: 3D mesh, domain Γ of vertices to which boundary conditions apply
  for each tetrahedron **do**
    compute initial volume V_0
    compute spatial derivatives of the shape functions δh
    compute linear strain-displacement matrices Z_0
  end for
  compute diagonal mass matrix M
  compute A_i, B_i and C_i (Eq. 2.27)

Algorithm 4 TLED algorithm: solver initialization.

Require: prescribed displacement $\mathbf{d}_{i\in\Gamma}(0)$
  initialize displacements: $\mathbf{u}(-\delta t) \leftarrow \mathbf{0}$, $\mathbf{u}(0) \leftarrow \mathbf{0}$
  initialize forces: $\mathbf{f}_i(-\delta t)$ and $\mathbf{f}_i(0)$
  initialize prescribed displacement $\mathbf{u}_{i\in\Gamma}(0) \leftarrow \mathbf{d}_{i\in\Gamma}(0)$

Algorithm 5 TLED algorithm: force computation at each iteration.

Require: $\mathbf{u}(t)$
  for each tetrahedron **do**
    compute deformation gradient $\mathrm{F}(t)$
    compute full strain-displacement matrix $\mathrm{Z}(t) \leftarrow \mathrm{Z}_0\mathrm{F}^T$
    compute T_a and T_p
    compute SPK tensor S at the integration points (Eq. 2.17)
    compute nodal internal forces: $\mathbf{f}(t) \leftarrow \int_{V_0} \mathrm{Z}(t)^T \mathrm{S}\, dV_0$
    add boundary conditions $\mathbf{f}(t) \leftarrow \mathbf{f}(t) + \mathbf{f}_{bp}(t) + \mathbf{f}_b(t)$
  end for

Algorithm 6 TLED algorithm: update step.

  for each node **do**
    update displacements $\mathbf{u}(u + \delta t)$ (Eq. 2.25)
    apply prescribed displacements: $\mathbf{u}_{i\in\Gamma}(t + \delta t) \leftarrow \mathbf{d}_{i\in\Gamma}(t + \delta t)$
  end for

simple deformation allows the computation of the analytical expression of the Cauchy stress as a function of the deformation gradient F. A convenient set up to study simple shear is given by a cube with the axes aligned to the fiber, sheet and normal vectors in the reference configuration. From this configuration, the deformation gradient and stress tensors can be calculated analytically. With $\mathbf{f}_0 = [1, 0, 0]^T$, $\mathbf{s}_0 = [0, 1, 0]^T$ and $\mathbf{n}_0 = [0, 0, 1]^T$, a shear γ in the fs plane, i.e. along the $\mathbf{f}_0$ direction, is described with the following deformation gradient (Fig. 2.24):

$$\mathrm{F} = \begin{bmatrix} 1 & \gamma & 0 \\ 0 & 1 & 0 \\ 0 & 0 & 1 \end{bmatrix} \tag{2.28}$$

resulting in the following

$$\mathrm{C} = \begin{bmatrix} 1 & \gamma & 0 \\ \gamma & \gamma^2+1 & 0 \\ 0 & 0 & 1 \end{bmatrix}, \quad \mathbf{f} = \mathbf{f}_0, \quad \mathbf{s} = \gamma\mathbf{f}_0 + \mathbf{s}_0, \quad \mathbf{n} = \mathbf{n}_0. \tag{2.29}$$

The invariants then write $I_1 = 3 + \gamma^2$, $I_4s = 1 + \gamma^2$ and $I_4f = I_4n = 1$, and the shear stress becomes:

$$\mathrm{T} = a\gamma \exp(b\gamma^2) + 2a_s\gamma^3 \exp(b_s\gamma^4) + a_{fs}\gamma \exp(b_{fs}\gamma^2) \tag{2.30}$$

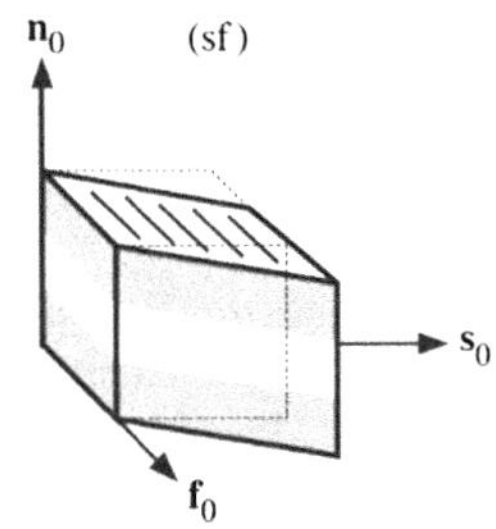

Figure 2.24. A mode of simple shear defined with respect to the fiber, sheet and normal axes. The first letter in (sf) stands for the normal vector to the face that is subject to shear, the second letter denotes the direction of shear. (Source: [120].)

Simple shear tests have also been conducted in vitro to characterize the material properties of the myocardium. The Holzapfel–Ogden model was in fact designed and fitted to available data [112] to reproduce the different stress-strain relationships that the myocardium exhibits in the three orthogonal planes. A verified implementation of the constitutive law should reproduce in silico the experimental results. Virtual tests of simple shear on a cube of material of finite dimensions (1 mm × 1 mm × 1 mm) were therefore simulated for verification. For each simple shear test, zero displacement on one face and a uniform displacement on the opposite face were prescribed, so as to apply a shear in each of the two orthogonal directions. We applied a finite shear in the range [0, 0.5] mm and computed the active shear stress. Finally, we compared the computed shear stress with available experimental data [112] as reported in [120]. The results are shown in Fig. 2.25, with a very good match between measured and computed values. All simulations were performed on a regular grid with size 0.1 mm, prescribing a time-dependent boundary condition so that the maximum level of shear was reached at $t = 0.5$ s. The time step was set to $\delta t = 10^{-5}$ s.

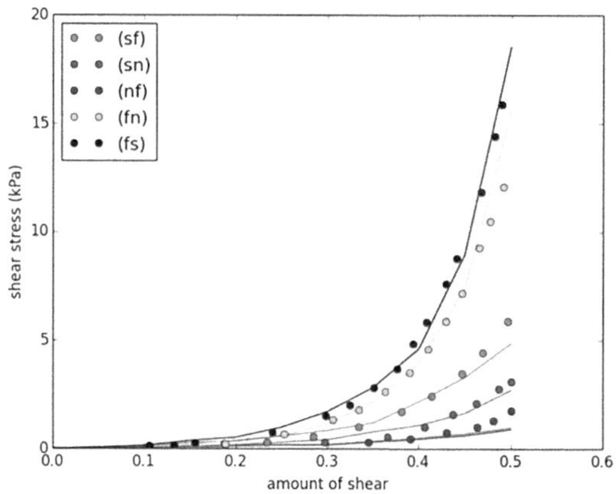

Figure 2.25. Simple shear tests on a finite sample of myocardial tissue. Circles represent experimental data as provided by [112]. Plain curves represent computational results. Each label summarizes the test as follows: the first letter stands for the normal vector to the face that is subject to shear, the second letter denotes the direction of shear.

Consistent with experimental observations and with the definition of the model, the numerical simulations predict different stress-strain relationships in the three planes. The accuracy of the computed stress was excellent in a wide range of shear, while decreasing for shear greater than 0.45. For high shear, in fact, the material tends to deform especially close to the loaded boundaries, and this causes the deformation gradient to change. Compression and traction components become more significant and the test is no longer a simple shear test. In these conditions, it would be important to verify that the assumptions of the model hold to replicate the experimental results (in particular regarding how the boundary conditions are prescribed in the experimental setup). For all experiments, the parameters reported in Table 1.1 were employed.

Numerical stability analysis

To verify the implementation of the TLED solver, a benchmark problem of linear elasticity was investigated, for which an analyt-

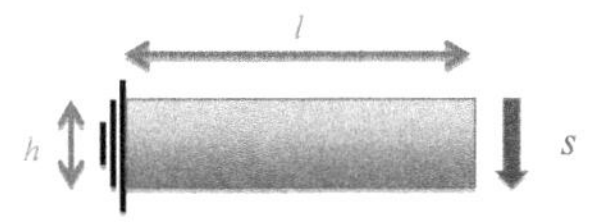

Figure 2.26. A linear elastic beam of length l and height h is subject to a suddenly applied shear stress S at one end while the other end is fixed.

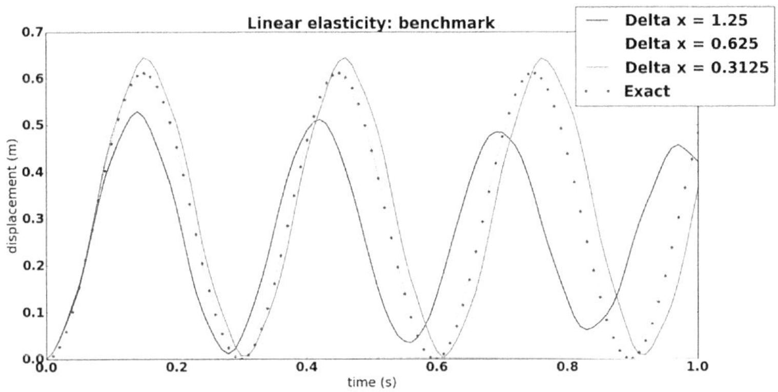

Figure 2.27. Displacement of the point at the center of the loaded boundary face, computed with different spatial resolutions and with the analytical solution of the problem.

ical solution was available. A beam of length $l = 20$ m and height $h = 5$ m was fully constrained at one end and subject to a suddenly applied constant shear stress $S = 1$ MPa at the other end (Fig. 2.26). All the other faces were traction free. A Young modulus of $E = 4$ MPa was used, with Poisson ratio of 0.32 and a structure density $\rho_s = 1450$ kg/m^3. The material reacted to the sudden shear load with a series of non damped sinusoidal oscillations about an equilibrium end displacement. The frequency ω and amplitude δ of the oscillations can be predicted by means of reduced models (1D beam theory) [228]: $\delta = 0.305$ m and $\omega = 3.35$ Hz.

The model was able to correctly describe the motion of the beam for a range of spatial discretizations (element sizes: 1.25 m, 0.625 m and 0.3125 m). The non damped oscillation of the beam was computed for 1 s, with time step $\delta t = 5 \times 10^{-5}$ s. The displacement of the point at the center of the boundary face subjected to the shear stress was observed and compared with the analytical solution. Fig. 2.27 shows that the numerical solution on the coarsest mesh suffered from significant numerical dissipation. On the two finer meshes, the solution was however in very good agreement with the analytical solution: the equilibrium displacement

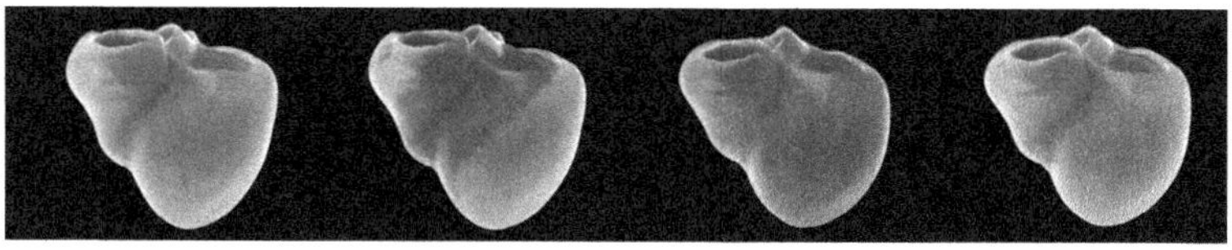

Figure 2.28. Example of bi-ventricular electromechanics simulation, from end-diastole to systole to relaxation. Color encodes the computed electrical potentials.

was captured in both cases with an error within 7%, while the predicted oscillation frequency was within 1% error with respect to the exact solution.

Bi-ventricular simulation

Finally, a typical result of a bi-ventricular simulation is presented in (Fig. 2.28). The geometry was extracted from cine MRI. Cuff pressure and 12-lead ECG were available to calibrate the model (see section 2.5). The behavior of the model in pathological conditions was then analysed by varying valve properties with various degrees of stenosis and regurgitations. The resulting pressure-volume loops are reported in Fig. 1.12, showing realistic changes in cardiac hemodynamics.

2.4 Hemodynamics modeling

The dynamics of blood is tightly coupled with the dynamics of the heart. On one hand, reduced order models of hemodynamics can be used to provide boundary conditions for the problem of cardiac motion. On the other hand, the motion of the heart determines complex flow patterns inside the heart chambers, which need to be captured with detailed dynamics models.

As discussed in section 1.4.1, reduced order models can be designed to describe various components of the circulatory system, including valves, arterial, atrial and venous circulation, and, where needed, the cardiac chambers. Implementation of numerical solvers for such models generally relies on standard techniques for the solution of ordinary differential equations and differential systems of algebraic equations. In the following we focus on the implementation and evaluation of computational methods for the full-order modeling of hemodynamics, discussing more in details some of the challenges and possible solutions.

2.4.1 3D hemodynamics using the lattice Boltzmann method

CFD modeling of intra-cardiac flow uses either unstructured, body-conforming grids (e.g. finite element methods [229–235] or finite volume methods [236–239]), or static, non-conforming grid (e.g. finite difference methods [131], the immersed boundary method [240–246], and recently also the Lattice Boltzmann method [247]). Given the significant domain distortion specific to cardiac simulation (wall and valve motion and deformation), any body-conformal grid method has to employ automatic remeshing strategies, which can significantly increase the complexity and cost of the simulation. In static grid methods the volume grid for the flow simulation does not need to be deformed or remeshed to conform to the deforming cardiac geometry, which makes them appropriate for the fully automated simulation of cardiac flow. The Immersed Boundary Method [248,249] is the oldest method that was used successfully to compute full 3D cardiac flow within complex moving geometries [250]. In contrast with finite volume methods, which impose the boundary conditions directly on the grid, the IBM introduces local body forces to achieve the same effect. While early on the method has been criticized for its inability to preserve mass accurately, later studies starting with [251] have shown that careful implementation can alleviate that. Furthermore, the Lattice Boltzmann method, whose efficiency in computing vessel blood flow has been established [247], is an interesting new option to explore, and it is considered here in more detail.

The Lattice Boltzmann Method

The Lattice Boltzmann Method (LBM) describes physics of fluid flow at a mesoscopic scale by taking into account molecular interactions between flow particles. The LBM provides ultimately the same solution as the Navier–Stokes based solvers [214], but it is "naturally" highly parallelizable, which can enable an efficient computation of two-way FSI. In the following, we provide a short description of the LBM theory, together with implementation details.

LBM models the interaction of fluid particles using a mathematical model based on the Boltzmann equation:

$$\frac{\partial f}{\partial t} + \mathbf{u} \cdot \nabla f = \mathrm{K}(f). \tag{2.31}$$

Here $f = f(\mathbf{u}, \mathbf{x}, t)$ is a probability density function and it gives the probability of a fluid particle to have the velocity $\mathbf{u}$ and to be at position $\mathbf{x}$ at time t. The right hand side of Eq. (2.18) is known as the collision operator and accounts for the contribution of the collision between particles. For the numerical implementation of LBM, Eq. (2.31) is written in a discrete form:

$$\frac{\partial f_i}{\partial t} + c_i \cdot \nabla f = \mathrm{K}(f_i), \tag{2.32}$$

where $f_i = f_i(\mathbf{x}, t)$ is a discrete representation of f with respect to the variable $\mathbf{u}$, more specifically instead of a single function f that depends on $\mathbf{u}$, $\mathbf{x}$, and t, there are a finite number of f_i functions that depend on just $\mathbf{x}$ and t. The discrete velocities c_i are associated to a lattice structure as displayed in Fig. 2.29, each velocity c_i corresponding to a link connecting a node $\mathbf{x}$ in the grid with a neighboring node $\mathbf{x} + \mathbf{e}_i$. The most commonly used lattice structures for 3D fluid computations contain 15, 19 or 27 links.

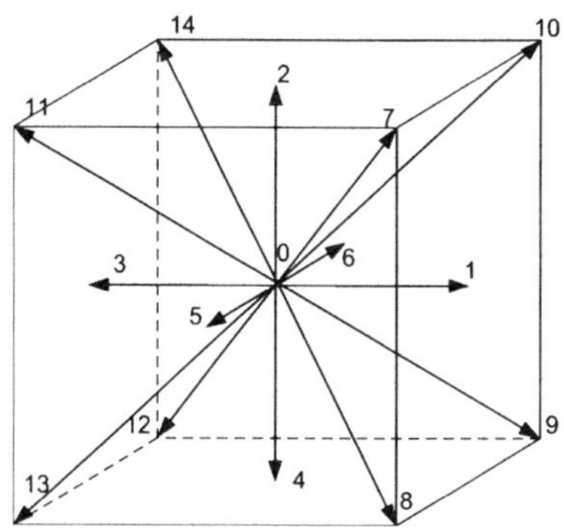

Figure 2.29. 15-velocity lattice structure.

The macroscopic pressure P and velocity $\mathbf{u}$ of the fluid are related to the density functions f_i as follows:

$$P = \rho c_s^2 = \sum_{i=0}^{N} f_i c_s^2, \tag{2.33}$$

$$\mathbf{u} = \frac{1}{\rho} \sum_{i=0}^{N} \mathbf{c_i} f_i, \tag{2.34}$$

where c_s is the non-dimensional speed of sound, equal to $1/\sqrt{3}$ for the 15, 19 and 27 velocity lattice structures.

Eq. (2.32) is solved using an explicit two-step time discretization scheme:

1. Collision: $f_i(\mathbf{x}, t + \Delta t) = f_i(\mathbf{x}, t) - \Omega_{i,j}(f_i(\mathbf{x}, t) - f_i^{eq}(\mathbf{x}, t))$
2. Propagation: $f_i(\mathbf{x} + \mathbf{c_i}, t + \Delta t) = f_i(\mathbf{x}, t)$.

The collision term $K_{i,j}(f_i(\mathbf{x},t) - f_i^{eq}(\mathbf{x},t))$ is formulated as a relaxation towards a thermodynamic equilibrium, where K is the collision matrix (or collision operator), containing the relaxation factors. A popular formulation is the multiple-relaxation-time (MRT) collision operator [252].

Turbulence modeling

Cardiac flow may enter physical regimes characterized by high Reynolds numbers, which require turbulence modeling to be added to the regular LBM description. The Smagorinsky model, for example, is a well established approach for simulating turbulent flows. It consists of modeling sub-grid scale effects as an additional viscosity ν_t, called turbulent viscosity. In our experience the Smagorinsky turbulence model, while simple to implement, was not stable enough for the high Reynolds numbers that can occur during cardiac flow physics computation. It should be mentioned that one of the most important limitations of the LBM is numerical instability that appears when viscosity is too low (or, equivalently, Reynolds number is too large).

An approach to improve the stability is the entropic Lattice-Boltzmann method (ELBM) [253], which is based on the Boltzmann's H theorem, and which consists of changing the collision model so that it satisfies the second law of thermodynamics. The equilibrium distribution function is changed as follows:

$$f^{eq}(\mathbf{x},t) = w_i \rho \prod_{d=1}^{D} \left(2 - \sqrt{1 + 3u_d^2}\right) \left(\frac{2u_d + \sqrt{1 + 3u_d}}{1 - u_d}\right)^{e_{id}}.$$

Accordingly, the collision operation becomes (with D the space dimension, 3 in our case):

$$f_i(\mathbf{x} + e_i \delta t, t + \delta t) = f_i(\mathbf{x},t) + \alpha\beta(f_i^{eq}(\mathbf{x},t) - f_i(\mathbf{x},t)),$$

where $\beta = 2/\tau$ is the equivalent of the LBGK relaxation time [252] and is related to the fluid viscosity. The additional factor α is specific to the ELBM model and it is computed for each lattice node at each iteration by solving a non-linear equation:

$$H(f) = H(f + \alpha(f^{eq} - f)), \tag{2.35}$$

$$H(f) = \sum_{i=0}^{N} f_i \ln\left(\frac{f_i}{w_i}\right). \tag{2.36}$$

Eq. (2.35) is usually solved iteratively using a Newton method. Note that if $\alpha = 2$ then the LBGK collision is recovered. In the

ELBM framework the simulation is stabilized by automatically tuning the α parameter, which allows the simulation to run at very high Reynolds numbers without becoming unstable or requiring large grid resolution.

2.4.2 3D fluid structure interaction

In this section we give specific implementation details for a Fluid Structure Interaction (FSI) cardiac computation system (see Fig. 2.30). As a reminder, the computation system is comprised of the TLED solver, handling the biomechanics computations, the 3D CFD solver, in this case LBM-CFD being used, and the valve module, which has two components, the reduced order one (part of TLED) and the 3D one. At any given time step of the computation the FSI algorithm proceeds as follows:

1. execute one step of TLED, updating the biomechanics (myocardial walls kinematics) and the 0D valve state (opening of the valve)
2. map 0D valve phase to the 3D valve kinematic phase, as constrained also by the myocardium (generating new 3D valve positions)
3. run 3D CFD using the new myocardial positions and velocities, and the new 3D valve positions and velocities
4. send 3D CFD pressures to myocardial walls and mean pressure to 0D valve modules.

The two main systems (TLED and LBM-CFD) run at different time steps, and the so-called "sub-cycling" approach ensures that information exchange is done at the appropriate time stamps. In practice, as TLED runs at coarser time steps (step 1 in the algorithm), the LBM-CFD solver is run for several time steps (the "sub-cycles", repeatedly executing step 3) with unchanged boundary conditions, until its time stamp gets in sync with TLED. We give more details in the following for each of the FSI steps.

Preparatory step

Some useful pre-computations need to occur first. The end-diastolic (at time $t = 0$) myocardial tetra mesh is provided with LV and LVOT tags that are used to extract an endocardial surface whose topology remains unchanged over the course of the cardiac cycle. The endocardial surface boundary edges are separated into the mitral and aortic contours, which are then tessellated to create the mitral and aortic virtual boundaries (outlets) for the use of the CFD code. Each endocardial mesh vertex is associated to a (closest at $t = 0$) tetra mesh vertex and follows rigidly its motion over the course of the cardiac cycle. As the endocardial mesh gets updated

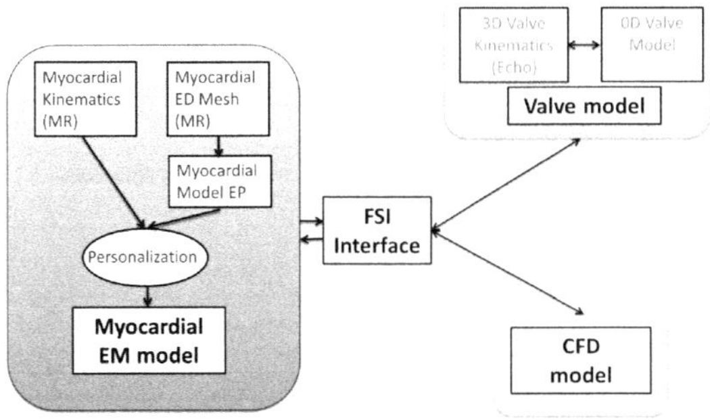

Figure 2.30. Fluid structure interaction system for cardiac haemodynamics computation. The interactions between the electromechanical model, valves and the CFD model are controlled by the FSI interface module.

at any time t, the aortic and mitral outlet surface barycenters are updated accordingly. The displacement rate of the endocardial vertices is the velocity information being sent to the CFD solver. Furthermore, each of the valve ring boundary vertices is kinematically linked to a closest myocardial tetra mesh vertex, computed at end-diastolic time ($t = 0$). During simulation, as the tetra mesh moves over time, the new tetra mesh position is imposed as a rigid constraint to the valve boundary vertices. This ensures that the valve boundary travels along with the mesh as it moves over time.

Another valve constraint is ensured using a 0D–3D kinematics mapping. As a reminder, the 0D valve model uses a valve phase function which is proportional to the effective area opening. In Fig. 2.31 we give an illustrative example of how the valve phases can be mapped to 3D valve mesh sequences over the course of the cardiac cycle, in this pre-computation stage. In practice one can start from a kinematic sequence of topologically consistent (i.e. satisfying point-to-point correspondence) 3D valve meshes (see section 2.1.1) and first compute their phases as the areas of the minimal surfaces spanning their rim boundaries. By focusing then on a given cardiac stage (e.g. systole), one can create a valve-mesh-over-systolic-time function by an interpolation method of choice. For example, a Fourier transform for each of the mesh nodes has also the benefit of smoothing any high frequency noise in the initial mesh sequence. The 0D–3D mapping is then used during the computation step to provide the 3D valve mesh corresponding to any 0D opening phase.

Step 1

This step has already been discussed in the biomechanics section. As a reminder (see section 1.4.1), the 0D valves use the ven-

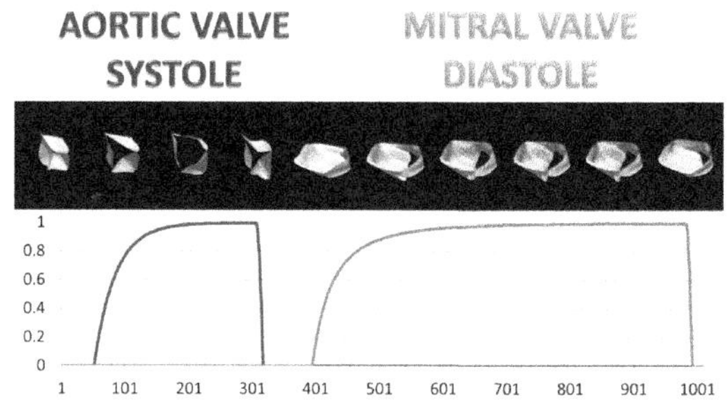

Figure 2.31. Aortic and mitral 3D valves are controlled by 0D opening phase functions whose dynamics is governed by pressure gradient forces.

tricular blood pressure and the atrial (or arterial) pressure to determine the opening phase of the valve (equivalent to the relative opening orifice area), which is a smooth function varying between 0 (closed) and 1 (open).

Step 2

During the second step two operations occur: first, the 3D valve configuration is obtained for the current opening phase. This operation can provide the kinematic mesh corresponding to any phase, however the challenge is to have such a mesh obey the boundary condition imposed by the dynamic myocardial mesh. This is done by the second operation.

Given the correspondence map between the valve base vertices and the myocardial mesh, the valve vertex positions are defined using a linear combination between a rigid transform (barycentric translation) and the transform given by the (myocardium constrained) base vertex kinematics. The two transforms are weighted using the relative distances from any valve vertex to its corresponding rim and base valve vertices. This ensures that the rim opening area/phase always matches the prescribed one from 0D, while the valve also follows the "live"/dynamic heart motion. A possible downside of this approach is that extra kinematic stretch is imposed on the valve base, which was found in practice to be insignificant nevertheless.

At the end of this step one obtains a 3D valve mesh whose base vertices lie on the myocardium with no gaps, while the orifice opening area corresponds to the one computed by the dynamic 0D valve model. The 3D valve position and velocity is then sent to the CFD solver.

Step 3

The third step runs the CFD solver using the new myocardial wall and 3D valve positions. Here we discuss in detail the treatment of the moving boundaries in LBM, which is of particular interest for us, given the complex deformations encountered by the endocardial wall and the cardiac valves. For lattice nodes located near a boundary, i.e. where there is a neighboring node located outside the fluid region, there are unknown f_i values that are required to complete the propagation step. The most commonly used way to compute the unknown distributions is the bounce-back approach: the unknown f_i values, propagated from a solid node, are set to the value corresponding to the opposite lattice direction. This is equivalent to reversing the velocity of a particle colliding with a static wall for example, $f_i = f_j$, where $\mathbf{c_i} = -\mathbf{c_j}$ are two opposite lattice directions. For moving and curved walls one can use the bounce-back scheme of [254]:

$$f_j(\mathbf{x}, t + \Delta t) = \begin{cases} 2q f_i(\mathbf{x}, t + \Delta t) + (1 - q) f_i(\mathbf{x} - \mathbf{c_i}, t + \Delta t) + 2w_i \mathbf{c_j} u_w \text{ if } q < 0.5 \\ \frac{1}{2q} f_i(\mathbf{x}, t + \Delta t) + \frac{2q-1}{2q} f_j(\mathbf{x}, t + \Delta t) + \frac{1}{q} w_i \mathbf{c_j} u_w \text{ if } q \geq 0.5 \end{cases}$$

where u_w is the wall velocity, and q is a factor between 0 and 1 that accounts for the exact position of the wall between two lattice nodes.

For the outflow boundary, i.e. wherever the flow leaves the domain, the velocity is usually unknown and the pressure is specified. In this case the non-equilibrium extrapolation method [255] is employed. The non-equilibrium extrapolation method replaces all the f_i values at the boundary using information extrapolated from neighboring locations:

$$f_j(\mathbf{x}, t + \Delta t) = f_i^{eq}(\mathbf{x}, t) + (1 - \Omega_{i,j}) f_i^{neq}(\mathbf{x_{neigh}}, t)$$

where $f_i^{neq} = f_i - f_i^{eq}$ is the non-equilibrium part of the distribution functions and x_{neigh} is a neighboring fluid node located along the boundary surface normal $\mathbf{n_b}$ such that $(\mathbf{x_{neigh}} - \mathbf{x}) \times \mathbf{n_b} = 0$. The f_i values at $\mathbf{x_{neigh}}$ are found by means of spatial interpolation.

Step 4

The fluid stress tensor definition is $\mathrm{T} = -pI + \mu(\nabla \mathbf{u} + \nabla \mathbf{u}^T)/2$, where p is the fluid pressure, $\mathbf{u}$ is the velocity and μ is the blood dynamic viscosity coefficient. Due to the relatively high Reynolds number regime of the cardiac flow and the high pressure loads, one can in a first approximation omit the contribution of the viscous stress, and approximate the stress tensor as $\sigma = -pI$. The

CFD relative pressure field passed on to the FSI interface is processed as follows:

- averaged inside the ventricle and subtracted from all points to ensure zero mean
- extrapolated across the endocardial surface using a second order accurate extrapolation method
- interpolated at the endocardial triangle barycenter locations.

This interpolated value, to which we add back the TLED pressure for re-gauging, is the absolute pressure field applied as a load on the endocardium.

Tests of the FSI module

We present here two simple verification tests of the FSI CFD model, which compare results of the presented FSI solver to analytical solutions.

Test 1: Peristaltic transport

The experiment consists of imposing a time-periodic, wave-like motion to a vessel geometry and in verifying the cross-sectional flow at any time step (as originally described in [256]). The wall deformation is changing both in time and space (along the axial direction) and the flow variation at every time step is computed analytically to depend on the wall movement as follows:

$$\Theta = \frac{3\phi^2}{2+\phi} \tag{2.37}$$

where $0 \leq \phi \leq 1$ represents the amplitude of the propagating wave, with $\phi = 0$ corresponding to total occlusion of the vessel and $\phi = 1$ corresponding to a fully open vessel profile. Fig. 2.32 outlines the favorable results obtained through numerical simulations. A total of five values for ϕ were used, with flow rates compared against the analytical solution. The experiments show excellent agreement with the analytical solution.

Test 2: Expanding and contracting vessel

This experiment validates the implementation of mass-conservation in the context of varying inlet and outlet boundary flows. The geometry was generated synthetically by applying a deformation function to a straight cylinder. It is given by the following equations:

$$\begin{aligned} x &= x \\ y &= R(x,t)\cos\theta \\ z &= R(x,t)\sin\theta \\ R(x,t) &= R_0 + R_{max}\sin(2\pi t)\sin(\pi x) \end{aligned}$$

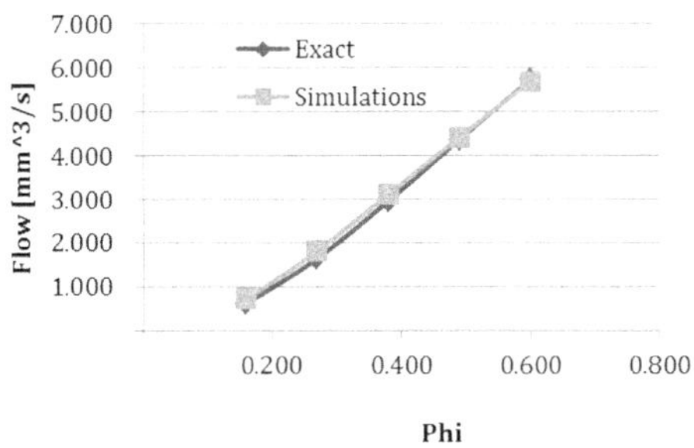

Figure 2.32. Cross-sectional flow variation with the peristaltic amplitude. An excellent match with theory is obtained.

where $R(x,t)$ is the vessel radius. In the simulations $R_0 = 0.5$ mm, $R_{max} = 0.3$ mm and the length of the vessel was 2 mm. The time period was one second (see Fig. 2.33.)

Figure 2.33. Time variation of the geometry of the expanding and contracting vessel.

Computations with the above described setup were performed and flow rates were compared with analytical solutions. Since the fluid is incompressible, the flow rate should be equal to the volume change. The volume change was calculated analytically from the surface equations describing the geometry. Inlet and outlet flow rates were computed on planes located at the inlet and outlet section, respectively. Results in Fig. 2.34 show that the simulated flow rates follow the analytical curve almost perfectly. Furthermore, the total flow rate (computed through a center cross section) was within 2 percent of the total analytical volume change.

Output of the FSI system with patient-specific data

In Fig. 2.35 we visualize results from a typical FSI computation using the presented algorithm, with cardiac geometry extracted from MRI images and valve geometry from 3D Ultrasound. Three-dimensional ventricular blood velocities and pressure fields, as well as myocardial stresses are available at all the points in the domain. Velocity fields display realistic patterns, including posterior jet deflection and mitral vortex appearance. The use of the kineto-

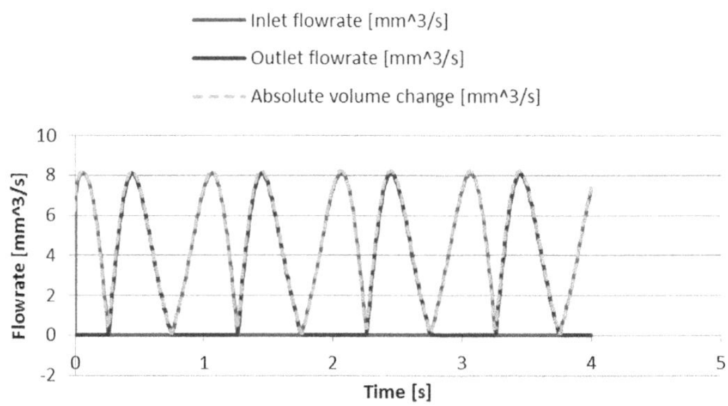

Figure 2.34. Flow rates vs time.

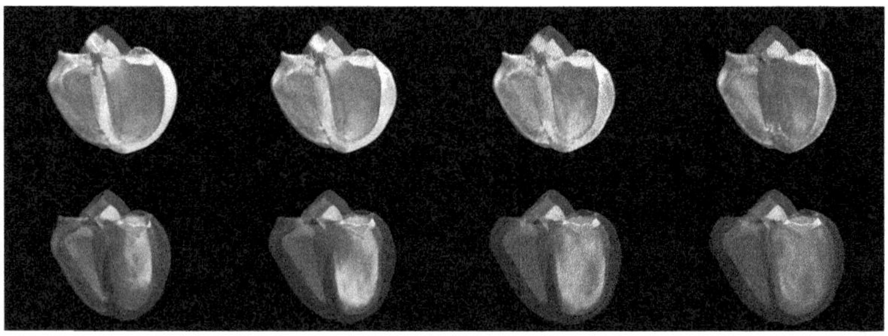

Figure 2.35. Cardiac cycle (systole on top, diastole on bottom) computed using the FSI framework introduced in this chapter. Velocity magnitude in the left ventricle is visualized using a standard rainbow colormap with constant positive slope transparency map. Myocardial stress magnitude is also visualized with a black-body radiation colormap.

dynamic valve system ensures that valve geometry at any time is also realistic.

2.5 Parameter estimation

The goal of model personalization is to estimate the free parameters of the model such that it captures the observed, clinically measured cardiac physiology. A successful strategy is to marginalize each modeling component and personalize them one-by-one, following a sequence that reflects their interdependency. For instance, one could adopt the following sequence:

1. Compute the patient-specific anatomical model of the heart from medical images (section 2.1).

2. Given the measured blood flow through the aortic and pulmonary valves, as well as pressure information, estimate the Windkessel parameters associated to the circulatory and pulmonary systems respectively (section 2.5.1).
3. Given electrophysiology data, typically 12-lead ECG traces, and the patient-specific anatomical model, estimate the electrophysiology parameters (section 2.5.2).
4. Given measured volume curves, pressure information, along with the personalized Windkessel and electrophysiology models, estimate the parameters of the biomechanical model (section 2.5.3).

The following sections describe the optimization techniques used to achieve each of these steps. AI-based methods for parameter estimations are presented in more details in chapter 5.

2.5.1 Windkessel parameters from pressure and volume data

As described in section 1.4.1, arteries are modeled using a Windkessel formulation. Model input is the blood through the vessels. Model output is the resulting pressure curve, controlled by the remote pressure, arterial compliance as well as characteristic and peripheral resistances. The model also has an initial condition to be estimated, namely the initial pressure.

Typically, one has at hand the flow curves through the arteries (e.g. obtained from color Doppler ultrasound or magnetic resonance imaging) and pressure information, either cuff pressure or invasive catheterization. Let us assume that continuous pressure and blood pool volume information are available. Both measurements are often acquired at different time points, with different heart rates and physiological conditions. Hence, a first step consists in temporally aligning the volume and pressure curves. This is achieved by first scaling the systolic portion of the pressure curve such that the ejection time observed through the pressure measurement matches the ejection time measured on the volume curve. The diastolic phase is then matched by temporal stretching (Fig. 2.36, left panel). Some low pass filtering may also be needed to reduce the noise in the measurements.

The model parameters are then estimated automatically by using the simplex method. Let $\mathbf{p}_m$ and $\mathbf{p}_c$ be the measured and computed arterial pressure respectively, N samples over one heart beat. An effective cost function to estimate all parameters but the initial pressure is:

$$\left[\min(\mathbf{p}_m) - \min(\mathbf{p}_c)\right]^2 + \left[\max(\mathbf{p}_m) - \max(\mathbf{p}_c)\right]^2$$

$$+\frac{1}{N}\sum_{i=1}^{N}(\mathbf{p}_m[i]-\mathbf{p}_c[i])^2 . \tag{2.38}$$

Finally, the initial pressure $p_c(0)$ is estimated by computing several heart cycles (Fig. 2.36, right panel). More precisely, $p_c(0)$ is adjusted automatically such that the first computed pressure cycle is as close as possible to the last one, which is assumed to have reached the periodic state.

2.5.2 Cardiac electrophysiology

The third step in the personalization procedure consists in fitting the cardiac electrophysiology model to the available data. In this section, 12-lead ECG features (QRS duration (QRSd), electrical axis (EA) and QT interval) are assumed to be available. The goal is therefore to estimate the conduction velocities of the myocardium (σ_{myo}), LV endocardium (σ_{LV}) and RV endocardium (σ_{RV}), as well as the action potential duration (τ_{close} in the case of the previously described LBM-EP model, or the APD value for Graph-EP).

To reduce the space of possible solutions, we rely on the physiological knowledge that depolarization in the Purkinje network is at least as fast as in normal myocytes, and about 2–4 times faster if healthy. As a result, the following constraint is imposed: $\sigma_{myo} < \sigma_{LV}$ and $\sigma_{myo} < \sigma_{RV}$. To simplify the notations, the model is identified by the function $f(\sigma_{myo}, \sigma_{LV}, \sigma_{RV}, APD)$. Parameter identification proceeds according to Algorithm 7. First, the parameters are initialized using values from the literature [42]. The

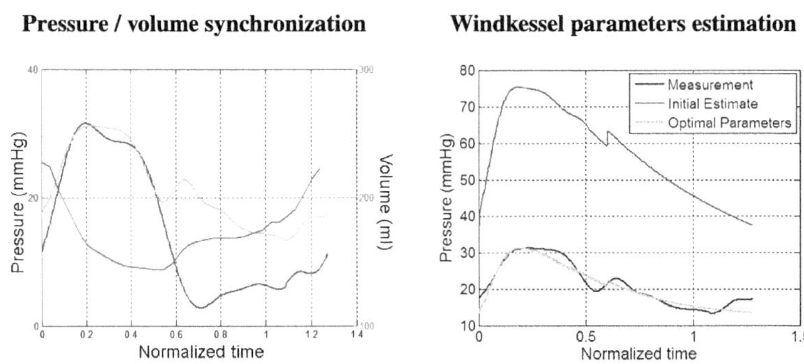

Figure 2.36. Different steps involved for the estimation of the Windkessel parameters of the arteries. Illustration on a pulmonary artery and right ventricular data. In the left panel, intra-cardiac pressure is shown in blue (dark gray in print version); arterial pressure in red (mid gray in print version) ; ventricular volume in green (light gray in print version) . Pressure is in mmHg, volume in ml.

parameter estimation process is then performed in two steps that are iterated N times, N being defined by the user. First, the global myocardium conductivity is estimated by matching the QRS duration. Then, the electrical axis is used to estimate the ratio between left and right conductivity. Both optimizations are performed using a gradient-free approach, for instance BOBYQA [187].

Algorithm 7 Cardiac electrophysiology personalization procedure from 12-lead ECG parameters.

Require: Get measured parameters $QRSd_m$, EA_m and QT_m
Require: Initialize σ_{myo}, σ_{LV}, σ_{RV} and APD with default parameters
Require: Set the number of iteration, typically $N \leftarrow 3$

function COMPUTEQRS(k, n)
  $(QRS_c, EA_c, QT_c) \leftarrow f\left(k \times (\sigma^n_{myo}, \sigma^n_{LV}, \sigma^n_{RV}), APD^n\right)$
  return QRS_c
end function

function COMPUTEEA(σ_{LV}, σ_{RV}, n)
  $(QRS_c, EA_c, QT_c) \leftarrow f\left(\sigma^{n+1}_{myo}, \sigma_{LV}, \sigma_{RV}, APD^n\right)$
  return EA_c
end function

for $i \leftarrow 1,\ i \leq N,\ i \leftarrow i+1$ **do**
  $k \leftarrow \text{argmin}_k(QRS_m - computeQRS(k, i-1))$
  $(\sigma^i_{myo}, \sigma^i_{LV}, \sigma^i_{RV}) \leftarrow k \times (\sigma^{i-1}_{myo}, \sigma^{i-1}_{LV}, \sigma^{i-1}_{RV})$
  $(\sigma^i_{LV}, \sigma^i_{RV}) = \text{argmin}_{(\sigma_{LV}, \sigma_{RV})}(EA_m - computeEA(\sigma_{LV}, \sigma_{RV}, i-1)$
  $(QRS_c, EA_c, QT_c) \leftarrow f(\sigma^i_{myo}, \sigma^i_{LV}, \sigma^i_{RV}, APD^{i-1})$
  $APD^i \leftarrow APD^{i-1} + (QT_m - QT_c)$
end for
return $\sigma^N_{myo}, \sigma^N_{LV}, \sigma^N_{RV}, APD^N$

This algorithm allows to model conduction pathway diseases, like left (right resp.) bundle branch block. In particular, the stimulation points of the left (right resp.) ventricle are first disabled to model the impaired bundle branches. Then, the hindered Purkinje network is modeled by forcing σ_{LV} (σ_{RV} resp.) to be smaller than $1.25 \times \sigma_{myo}$. It should be noted that scars and fibrosis can easily be integrated by setting default values to these areas. The values could also be personalized by adding one additional step in Algorithm 7, following the same pattern.

2.5.3 Myocardium stiffness and maximum active stress from images

The last step of the personalization workflow is to estimate the biomechanical model parameters. As illustrated in Fig. 2.37, the personalized Windkessel and electrophysiology models are given as input, along with the clinical data used as reference. In brief, a gradient-free optimization method is used, for instance BOBYQA [187], to minimize a cost function that compares the observed cardiac dynamics (from images and hemodynamics parameters) with the computed values.

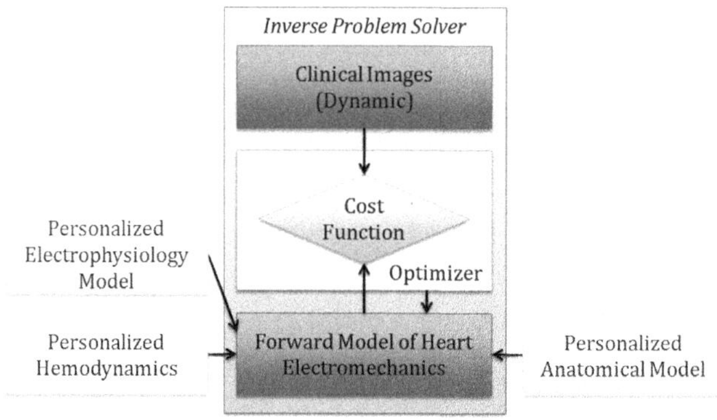

Figure 2.37. Inverse problem framework for personalizing the biomechanical model parameters from clinical data.

Using the already personalized electrophysiology and Windkessel models, one can generate the forward simulations necessary to the optimization algorithm. The key is to define a representative feature set from all the hemodynamics and kinematics parameters that can be computed to form an effective cost function to minimize. In particular, the features need to be representative enough and, if possible, as independent as possible, to ensure maximal parameter observability while minimizing risks of confusion due to data noise.

Let $V(t)$ and $P(t)$ be the time-varying ventricular volume and pressure curves. In this example, the feature set $\boldsymbol{\Omega}$ is composed of six hemodynamics features per ventricle:

Ω_1 : stroke volume $SV = \max V(t) - \min V(t)$
Ω_2 : ejection fraction $EF = SV/\max V(t)$
Ω_3 : minimum volume $\min V(t)$
Ω_4 : maximum pressure $\max P(t)$
Ω_5 : volume curve $\mathbf{V}$ (sampled vector of $V(t)$)
Ω_6 : pressure curve $\mathbf{P}$ (sampled vector of $P(t)$).

The parameters to estimate are the maximum active stress τ_0, the stiffness coefficient β, the contraction and relaxation rates k_{ATP} and k_{RS} respectively, for both left and right ventricles. They form the parameter vector $\boldsymbol{\theta} = (\tau_{0,LV}, \beta_{LV}, k_{ATP,LV}, k_{RS,LV}, \tau_{0,RV}, \beta_{RV}, k_{ATP,RV}, k_{RS,RV})$. The parameters are first initialized to default values obtained from the literature, and then optimized by minimizing the cost function ψ:

$$\psi(\boldsymbol{\Omega}_m, \boldsymbol{\Omega}_c) = \sum_{i=1..6} \lambda_i \cdot \mathcal{D}(\Omega_{m,i}, \Omega_{c,i}) \tag{2.39}$$

where the indices m and c denote "measured" and "computed" respectively. $\mathcal{D}$ is a distance function. For scalar parameters, $\mathcal{D}(a, b) = (a - b)^2$. For vectorial parameters, $\mathcal{D}(\mathbf{a}, \mathbf{b}) = ||\mathbf{a} - \mathbf{b}||_{\mathcal{L}_2}$, the $\mathcal{L}_2$ norm. The λ_i are weighting coefficients, with $\boldsymbol{\lambda} = (3, 2, 1, 1, 2, 2, 3, 2, 1, 1, 2, 2)$ to improve optimization performance by increasing the weights of the most reliable features, as observed experimentally.

The pulmonary vein pressure is also estimated during the personalization of the biomechanical model. This is achieved in two steps. First, an initial computation is performed with generic parameters prior to biomechanical personalization. The pulmonary vein parameter of the model is adjusted by adding the difference between measured and computed atrial pressure at beginning of diastole. After biomechanical personalization, the pulmonary vein is further adjusted if needed.

It is worth noting the cost function must be calculated after three or more heart cycles to minimize the effects of the transient regime. As a result, estimating biomechanical parameters can quickly become time consuming. Even if one heart cycle takes only 2–3 minutes to compute, personalizing a cardiac model could still take few hours to converge due to the large number of iterations needed by traditional gradient-free optimization methods. It is therefore crucial to develop computational methods that allow either fast simulation or fast parameter estimation, or both.

2.6 Summary

Nowadays, modelers and computer scientists can leverage a wide variety of numerical discretization schemes to solve the complex equations that govern the laws of cardiac function. This chapter presents a specific implementation strategy, targeting fast, multi-scale and personalized modeling. Other methodologies are possible of course, as it can be seen in the literature. Some approaches are more optimized for computer graphics, favoring computational efficiency and realism over physiological accuracy, whereas other methods aim at achieving high numerical accuracy at the expanse of computational burden. With the development of AI, new methods are being investigated to enable fast, personalized simulations of cardiac function. As the reader will see in the second part of the book, these new approaches will likely be cornerstone of next generation, real-time heart modeling applications.

2

Artificial intelligence methods for cardiac modeling

3

Learning cardiac anatomy

Florin C. Ghesu, Bogdan Georgescu, Yue Zhang, Sasa Grbic, Dorin Comaniciu
Siemens Healthineers, Princeton, NJ, United States

3.1 Introduction

Parsing of cardiac and vascular structures is an essential prerequisite for cardiac image analysis and represents the first step in constructing a comprehensive heart model. In practice, parsing refers to the detection, segmentation and tracking of anatomical structures. This information enables not only the direct derivation of measurements (distances, volumes, ejection fraction, etc.) for quantitative assessment, but also the estimation of physiological heart models or the mechanical simulation of blood flow. The features extracted during the parsing step are often critical for the timely diagnosis of acute conditions, serving as input for artificial intelligence methods for diagnosis or interventional guidance.

In this chapter, we present several state-of-the-art methods for cardiac image parsing. We cover the marginal space learning [31] and marginal space deep learning [257,258] frameworks, demonstrating their performance at detecting and segmenting various structures, such as heart chambers and valves based on 3D computed tomography (CT) and 3D ultrasound (US) images. We also present the concept of multi-scale image navigation, as an efficient alternative to exhaustive image scanning. The accuracy and speed of this new framework at detecting various cardiac landmarks, as well as vascular landmarks in the body, are reported on several 2D magnetic resonance (MR), 2D-US and 3D-CT image datasets. We also present a modern deep image-to-image fully convolutional segmentation network and report its performance at segmenting the entire heart. Finally, we discuss the structure tracking problem in cardiac modeling and review the state-of-the-art deep learning based methods.

Artificial Intelligence for Computational Modeling of the Heart
https://doi.org/10.1016/B978-0-12-817594-1.00014-0

3.2 Parsing of cardiac and vascular structures

3.2.1 From shallow to deep marginal space learning

3.2.1.1 Problem formulation

Let us reformulate object localization as a classification problem. Boxes of image intensities, i.e., constrained axis-aligned subregions of the image, parameterized as $h \in \mathcal{U}$ are sampled from the image. The variable $\mathcal{U}$ denotes the parameter space. We distinguish between positive samples (centered around the ground-truth object position), and negative samples which are extracted from the rest of the image. A classifier is trained to distinguish between these two categories. At test time, the classifier is applied exhaustively to scan the complete parameter space and yield the most probable positions of an object of interest. In practice, however, this strategy is subject to severe computational limitations, as the scanning effort grows exponentially with the dimensionality of the parameter space. Given an arbitrary anatomical object of interest, the goal is to estimate its restricted affine transformation which is defined by nine parameters. They define the location $\mathbf{T} = (t_x, t_y, t_z)$, the orientation $\mathbf{R} = (\phi_x, \phi_y, \phi_z)$, and the anisotropic scale $\mathbf{S} = (s_x, s_y, s_z)$ of the considered object. One can observe that a simple coarse discretization of $d = 10$ possible values per parameter results in highly prohibitive number of $d^9 = 1,000,000,000$ hypotheses that need to be evaluated by the classifier. Marginal Space Learning and its deep learning based extension aim at drastically reducing this computational complexity by restricting the portion of the parameter space in which candidates are searched.

Given a bounding box computed with marginal space deep learning (MSDL), one can use this information to estimate the nonrigid deformation of the object. An initial estimation of the shape is obtained by rigidly transforming the mean object shape according to the estimated pose. This initial estimate is then iteratively refined using an active shape model based on deep learned image features. In more detail, a boundary classifier is trained to decide whether there is a boundary point at a given position and under a given orientation. To solve this problem, Zheng et al. [31] proposed steerable features combined with the PBT [203].

Once the pose of the object has been estimated, its shape needs to be determined. Statistical shape modeling is used to obtain a parametric description of the nonrigid deformation of the object, where $c_1, c_2, \ldots, c_K \in \mathbb{R}$ are the coefficients of the major deformation modes, with K selected so that the object shape can be closely approximated. Using a boundary classifier that is trained to distinguish points on the object surface from points that are not on the surface, one can estimate the K coefficients at runtime and thereby obtain a surface segmentation of the object.

3.2.1.2 Traditional feature engineering

Based on this formulation, Zheng et al. [31] proposed to use the Probabilistic Boosting Tree (PBT) [203] as discriminative learner. In terms of feature computation, 3D Haar wavelets [259] were extracted and used to encode the image information for translation estimation. Haar wavelets can be computed very efficiently and can easily generalize to high dimensions. However, their application is limited for capturing orientation and scale information. The extension requires a pre-alignment of the volume and the wavelet sampling pattern, which is very tedious and time-consuming for a 3D learning problem.

A fast alternative, that comes at the expense of missing global information, is the selection of local image intensity features. In this context, steerable features were proposed [31]. The idea of steerable features is to use a flexible sampling pattern to determine the image points at which local features are computed. For a given hypothesis $(x, y, z, \phi_x, \phi_y, \phi_z, s_x, s_y, s_z)$ the sampling pattern is centered at position (x, y, z), rotated by the corresponding angles (ϕ_x, ϕ_y, ϕ_z) and anisotropically scaled with the factors (s_x, s_y, s_z). Assuming that N local features are computed over a pattern of P sampling points, the complete feature pool will contain $P \times N$ features. With this strategy, Zheng et al. [31] demonstrate that one can effectively capture both global and local information and by steering the pattern, also incorporate orientation and scale information.

3.2.1.3 Sparse adaptive deep neural networks

An alternative to handcrafted features was proposed in [258]. The proposed classifier is based on a modern deep neural network architecture that supports the implicit learning of image features for image classification directly from the raw image signal. The application of standard deep neural network architectures is not feasible in the volumetric setting, mainly due to the complexity of the sampling operation under the considered object transformations. To address this challenge, Ghesu et al. [258] propose to enforce sparsity in the network architecture to significantly accelerate the sampling operation. The derived architecture is called: sparse adaptive deep neural networks, or SADNNs (see Fig. 3.1).

Given a fully-connected network architecture, the aim is to find a sparsity map $\mathbf{s}$ for the network weights $\mathbf{w}$, such that over T train-

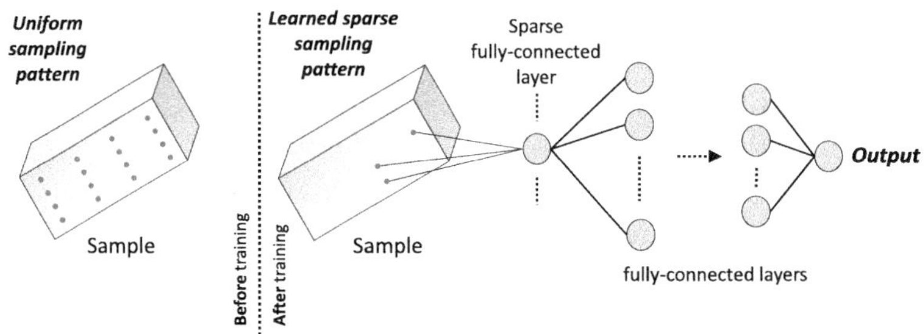

Figure 3.1. Visualization of uniform feature patterns versus self-learned, sparse, adaptive patterns.

ing rounds, the response residual ϵ given by:

$$\epsilon = \|\mathcal{R}(\mathbf{X}; \mathbf{w}_s, \mathbf{b}_s) - \mathbf{y}\|_2^2 \tag{3.1}$$

is minimal, where $\mathbf{b}_s$ are the biases of neurons in the sparse network, $\mathbf{w}_s$ are the learned sparse weights, determined by the sparsity map $\mathbf{s}$ with $s_i \in \{0, 1\}, \forall i$, $\mathcal{R}$ denotes to network response function, $\mathbf{X}$ the input image training data and $\mathbf{y}$ the corresponding reference $\{0, 1\}$ classification flags. In a greedy learning strategy, neural connections with minimal impact on the network response function are gradually eliminated, while continuing the training on the remaining active connections (see Algorithm 8). In each round $t \leq T$, this reduces to a subset of neural network connections with minimal absolute value that are selected and removed from the network. The L_1-norm (see Algorithm 8) is used to normalize the filter after each sparsity enforcement step. The training is continued on the remaining active connections, allowing the remaining neurons to adapt to the missing information (see step 12 of Algorithm 8):

$$\left(\hat{\mathbf{w}}^{(t)}, \hat{\mathbf{b}}^{(t)}\right) = \arg \min_{\substack{\mathbf{w}: \mathbf{w}^{(t)} \\ \mathbf{b}: \mathbf{b}^{(t)}}} \|\mathcal{R}(\mathbf{X}; \mathbf{w}, \mathbf{b}) - \mathbf{y}\|_2^2, \tag{3.2}$$

where $\mathbf{w}^{(t)}$ and $\mathbf{b}^{(t)}$ (computed from the values in round $t - 1$) are used as initial values in the optimization step. For more details on the methodology, please refer to [257].

Sparse adaptive data sampling patterns are learned, focusing the attention of the network on the most relevant information in the image and explicitly disregarding input with minimal impact on the network response function $\mathcal{R}$ (see Fig. 3.1). The experiments demonstrate that the sparse patterns can reach sparsity levels of **90–95%**. There are several benefits of this learning strategy: first, the sampling efficiency is increased by around 2 orders

Algorithm 8 Learning algorithm with iterative threshold-enforced sparsity.

Pre-training state: $\mathbf{w}^{(0)} \leftarrow \mathbf{w}$ (small number of epochs)
Initialize sparsity map $\mathbf{s}^{(0)} \leftarrow \mathbf{1}$
$t \leftarrow 1$
for each training round $t \leq T$ **do**
  for all filters i with sparsity **do**
    $\mathbf{s}_i^{(t)} \leftarrow \mathbf{s}_i^{(t-1)}$
    Update sparsity map $\mathbf{s}_i^{(t)}$ (remove smallest active weights)
    $\mathbf{w}_i^{(t)} = \mathbf{w}_i^{(t-1)} \odot \mathbf{s}_i^{(t)}$
    Normalize active coefficients s.t. $\|\mathbf{w}_i^{(t)}\|_1 = \|\mathbf{w}_i^{(t-1)}\|_1$
  end for
  $\mathbf{b}^{(t)} \leftarrow \mathbf{b}^{(t-1)}$
  Train network on active weights (small number of epochs)
  $t \leftarrow t + 1$
end for
Output sparse kernels: $\mathbf{w}_s \leftarrow \mathbf{w}^{(T)}$
Output bias values: $\mathbf{b}_s \leftarrow \mathbf{b}^{(T)}$

of magnitude; and second, the accuracy of the model is improved by the regularization effect of the imposed sparsity.

3.2.1.4 Marginal space deep learning

Given an observed input image $\mathbf{I}$, the estimation of the transformation parameters is equivalent to maximizing the posterior probability:

$$\left(\hat{\mathbf{T}}, \hat{\mathbf{R}}, \hat{\mathbf{S}}\right) = \arg\max_{\mathbf{T},\mathbf{R},\mathbf{S}} p(\mathbf{T}, \mathbf{R}, \mathbf{S}|\mathbf{I}). \tag{3.3}$$

Considering the large dimensionality of this parameter space, scanning becomes infeasible. In the marginal space learning (MSL) framework, Zheng et al. [31] propose to split this space in a hierarchy of clustered, high-probability subspaces of increasing dimensionality. They distinguish between the position space, the position-orientation space and the full 9D space including also the anisotropic scaling information of the object. This space separation is based on the following factorization of the posterior:

$$\begin{aligned}\left(\hat{\mathbf{T}}, \hat{\mathbf{R}}, \hat{\mathbf{S}}\right) &= \arg\max_{\mathbf{T},\mathbf{R},\mathbf{S}} p(\mathbf{T}|\mathbf{I})p(\mathbf{R}|\mathbf{T};\mathbf{I})p(\mathbf{S}|\mathbf{T},\mathbf{R};\mathbf{I}) \\ &= \arg\max_{\mathbf{T},\mathbf{R},\mathbf{S}} p(\mathbf{T}|\mathbf{I})\frac{p(\mathbf{T},\mathbf{R}|\mathbf{I})}{p(\mathbf{T}|\mathbf{I})}\frac{p(\mathbf{T},\mathbf{R},\mathbf{S}|\mathbf{I})}{p(\mathbf{T},\mathbf{R}|\mathbf{I})},\end{aligned} \tag{3.4}$$

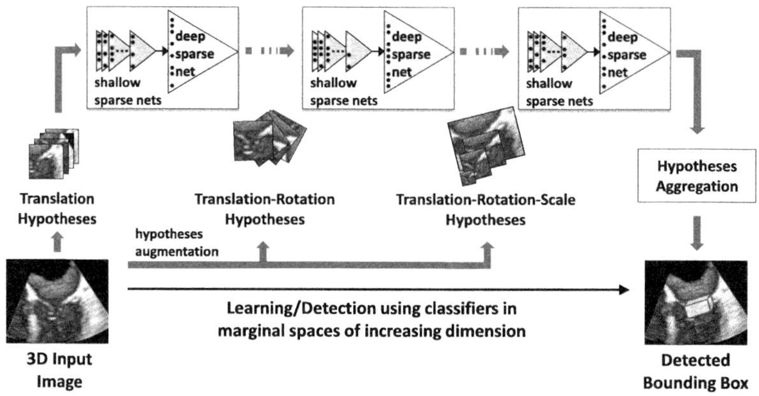

Figure 3.2. Schematic visualization of the marginal space deep learning framework applied for the sake of example to object localization in 3D echocardiographic images. The same approach can be used to parse images from different imaging modalities.

where the probabilities $p(\mathbf{T}|\mathbf{I})$, $p(\mathbf{T},\mathbf{R}|\mathbf{I})$ and $p(\mathbf{T},\mathbf{R},\mathbf{S}|\mathbf{I})$ are defined in the marginal spaces (see Fig. 3.2). We approximate each of these distributions with a SADNN: $\mathcal{R}(\mathbf{X};\mathbf{w}_s,\mathbf{b}_s)$. The first step is the learning of the translation parameters in the translation space $\mathcal{U}_T(\mathbf{I})$. The positive hypotheses with highest probability, clustered in a dense region, are augmented with discretized orientation information, to define the translation-orientation space $\mathcal{U}_{TR}(\mathbf{I})$. Similarly one can extend to the complete 9D space $\mathcal{U}_{TRS}(\mathbf{I})$. The optimization is defined:

$$
\begin{aligned}
\left(\hat{\mathbf{T}},\mathcal{U}_{TR}(\mathbf{I})\right) &\leftarrow \arg\max_{\mathbf{T}} \mathcal{R}\left(\mathcal{U}_T(\mathbf{I});\mathbf{w}_s,\mathbf{b}_s\right)\\
\left(\hat{\mathbf{T}},\hat{\mathbf{R}},\mathcal{U}_{TRS}(\mathbf{I})\right) &\leftarrow \arg\max_{\mathbf{T},\mathbf{R}} \mathcal{R}\left(\mathcal{U}_{TR}(\mathbf{I});\mathbf{w}_s,\mathbf{b}_s\right)\\
\left(\hat{\mathbf{T}},\hat{\mathbf{R}},\hat{\mathbf{S}}\right) &\leftarrow \arg\max_{\mathbf{T},\mathbf{R},\mathbf{S}} \mathcal{R}\left(\mathcal{U}_{TRS}(\mathbf{I});\mathbf{w}_s,\mathbf{b}_s\right),
\end{aligned}
\tag{3.5}
$$

where $\mathcal{R}(\cdot\,;\mathbf{w}_s,\mathbf{b}_s)$ denotes the response of each of the sparse adaptive deep neural networks, learned from the supervised training data in each marginal space. Using this type of hierarchical parameterization brings a speed-up of 6 orders of magnitude compared to the exhaustive search (see proof in [31]). In addition, to cope with the large number of negative hypotheses, a cascaded filtering procedure is proposed in [257,258,260]. Within this cascade, shallow sparse adaptive neural network models are trained to hierarchically reject negative hypotheses in an efficient way. The complete pipeline is visualized in Fig. 3.2.

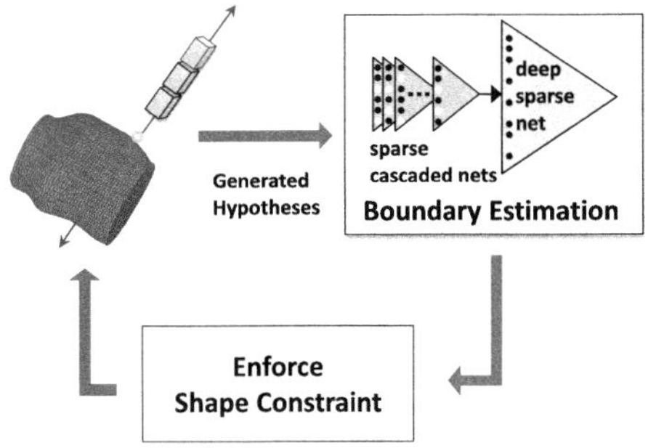

Figure 3.3. Schematic visualization of the learning-based boundary deformation step.

3.2.1.5 Nonrigid parametric deformation estimation

Given a bounding box computed with marginal space deep learning (MSDL), one can use this information to estimate the nonrigid deformation of the object. An initial estimation of the shape is obtained by rigidly transforming the mean object shape according to the estimated pose. This initial estimate is then iteratively refined using an active shape model based on deep learned image features. In more detail, a boundary classifier is trained to decide whether there is a boundary point at a given position and under a given orientation. To solve this problem, Zheng et al. [31] proposed steerable features combined with the PBT [203].

An effective alternative is to use cascaded SADNNs to learn adaptive, sparse feature sampling patterns around the boundary. These classifiers are trained to answer whether there is a boundary point at a given position $\mathbf{T} = \left[t_x, t_y, t_z\right]$ and orientation $\mathbf{R} = \left[\phi_x, \phi_y, \phi_z\right]$ on the warping shape. The orientation is defined by the normal for the respective shape point. The training is performed with positive samples on the ground-truth boundary of training examples (aligned with the shape normal) and negative samples at different higher distances from the boundary. The sparse adaptive patterns are essential in efficiently applying this classifier under arbitrary orientations.

Fig. 3.3 shows the iterative algorithm proposed for segmentation. After each boundary estimation step, the deformed shape is constrained to a subspace of shapes. Statistical shape modeling is used for the constraint. The boundary estimation step and shape constraint enforcement are applied in an iterative manner until convergence.

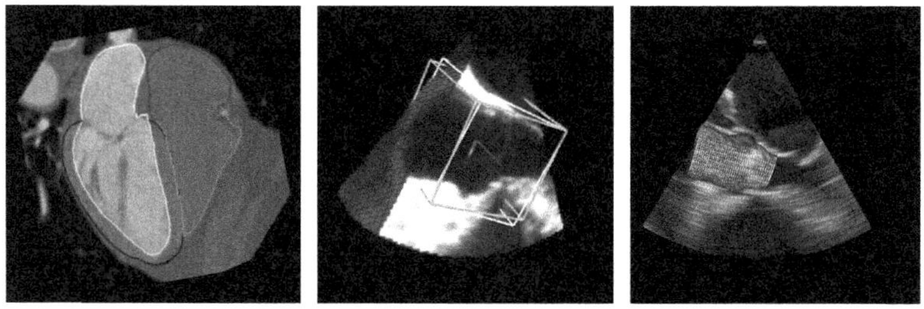

Figure 3.4. From left to right: segmentation results for all four heart chambers visualized in an orthogonal slice-view of a cardiac CT image (computed using MSL); a detected bounding box around the aortic heart valve in a 3D TEE Ultrasound volume (the detected box is displayed in green (mid gray in print version) and the ground-truth in yellow (light gray in print version); and finally, a 3D rendering of the corresponding triangulated surface mesh for the aortic root (both computed using MSDL).

3.2.1.6 Experiments

We present experiments on two datasets. First, we highlight the performance of the MSL solution on detecting and segmenting the heart chambers from cardiac 3D-CT data. The focus of this section is then set on comparing the traditional MSL to the redefined MSDL approach on detecting and segmenting the aortic valve (root) in 3D transesophageal echocardiogram (TEE) images.

The cardiac 3D-CT data contains 323 volumes with all four heart chambers manually annotated by experts. An additional 134 volumes and annotations were considered for the left ventricle. Based on a four-fold cross-validation experiment, the MSL method based on PBT classifiers and steerable features achieved a mean segmentation error between 1.21–1.57 mm (standard deviation below 0.5 mm) for all heart chambers. In particular, no outliers have been observed. More details on the algorithm setup and configuration, as well as validation details can be found in [31].

The cardiac 3D-US dataset contains 2891 3D TEE images of the heart from 869 patients. All images were resampled to an isotropic resolution of 3 mm. The validation is based on a random patient-based split of the volumes in 2481 training and 410 test volumes. Ground-truth was obtained through annotation by radiologists. For a complete definition of the aortic model, please see [32].

The network architecture is defined as follows: shallow model for the cascade with 2 layers – 5832 (sparse) $\times$ 60 $\times$ 1 hidden units and main SADNN with 4 layers – 5832 (sparse) $\times$ 150 $\times$ 80 $\times$ 50 $\times$ 1 hidden units. The input patch size is set to 18 $\times$ 18 $\times$ 18 voxels.

We quantify the accuracy of the localization using two measures: the distance between the box centers and a corner distance

Table 3.1 Result comparison for aortic valve detection in 3D US. Superior results are displayed in bold.

	Position error [mm]				Corner error [mm]			
	Training data		Test data		Training data		Test data	
	MSL	MSDL	MSL	MSDL	MSL	MSDL	MSL	MSDL
Mean	3.12	**1.47**	3.34	**1.83**	5.42	**2.80**	6.16	**3.72**
Median	2.80	**1.27**	3.05	**1.58**	4.98	**2.58**	5.85	**3.34**
STD	1.91	**0.99**	1.85	**1.31**	2.47	**1.23**	2.31	**1.74**

Table 3.2 Result comparison for aortic valve segmentation in 3D-US. Superior results are displayed in bold.

	Mean mesh error [mm]			
	Training data		Test data	
	MSL	MSDL	MSL	MSDL
Mean	1.04	**0.89**	1.17	**0.90**
Median	0.98	**0.82**	1.05	**0.80**
STD	0.50	**0.35**	0.66	**0.48**

error, i.e., the average distance between the 8 corners of the detected box and the ground-truth box. The second measure captures also the orientation and scale variability. Results are shown in Table 3.1. The MSDL framework shows a superior performance to the MSL solution, reducing the error by at least 40% with respect to both measures at runtime of under 0.5 seconds per case on CPU. An example detection is shown in Fig. 3.4.

The segmentation accuracy is measured by the average distance between the final mesh and the ground-truth mesh. As can be seen in Table 3.2, the MSDL approach outperforms the reference method [31] on average by 23%. Fig. 3.4 shows qualitative results.

3.2.2 Intelligent agent-driven image parsing

An exhaustive scanning strategy for estimating transformation parameters remains suboptimal, even after reducing the scanning effort to parameter subspaces via marginal space deep learning. In particular, the scanning process can become very time-consuming on high-resolution volumetric data. To address this challenge, we propose to reformulate the anatomical object detection problem as a search task for an artificial agent. Using elements of reinforcement learning and deep learning [261], we design an algorithm to teach artificial agents optimal navigation trajectories through the image space towards the anatomical structures of interest [262–265].

3.2.2.1 Learning to search for anatomical objects

In contrast to exhaustive search, an artificial agent learns how to find a structure or landmark by navigating through the space of a given image $\mathbf{I}: \mathbb{Z}^3 \rightarrow \mathbb{R}$. A Markov Decision Process (MDP) [266] models the dynamics of the navigation: $\mathcal{M} := (\mathcal{S}, \mathcal{A}, \mathcal{T}, \mathcal{F}, \gamma)$. The components are as follows: $\mathcal{S}$ is a finite set of states, $s_t \in \mathcal{S}$ being the state of the agent at time t (we define this based on the image context around the position of the agent at time t); $\mathcal{A}$ is a finite set of actions for voxel-wise navigation, i.e., ± 1 voxels along each image axis; $\mathcal{T}: \mathcal{S} \times \mathcal{A} \times \mathcal{S} \rightarrow [0, 1]$ is the stochastic transition function; $\mathcal{F}: \mathcal{S} \times \mathcal{A} \times \mathcal{S} \rightarrow \mathbb{R}$ is the reward (feedback) function used as incentive for behavior learning ($\mathcal{F}_{s,a}^{s'} = \|\mathbf{p}_t - \mathbf{p}_{GT}\|_2^2 - \|\mathbf{p}_{t+1} - \mathbf{p}_{GT}\|_2^2$ defines the expected distance-based reward for transitioning from state s to state s', i.e., from point $\mathbf{p}_t$ to $\mathbf{p}_{t+1}$ while searching for the location of the target structure); and $\gamma \in (0, 1)$ is the discount-factor, balancing immediate and future rewards [265].

Based on these components, we define the optimal action-value function $Q^*: \mathcal{S} \times \mathcal{A} \rightarrow \mathbb{R}$ that measures the maximum expected discounted future reward R_t of an optimal navigation policy π^*: $Q^*(s, a) = \max_\pi \mathbb{E}[R_t | s_t = s, a_t = a, \pi]$. One can derive a recursive form of this function, also referred to as the Bellman optimality criterion [266]: $Q^*(s, a) = \mathbb{E}_{s'}\left(r + \gamma \max_{a'} Q^*(s', a')\right)$ (r is the reward from s to s'). We propose to use a parametric model in the form of a deep neural network to approximate this function: $Q^*(s, a) \approx Q(s, a; \theta)$ (θ defines the parameters). Based on Q-learning [267,268], one can learn an effective image navigation strategy for finding anatomical structures with maximum reward [261,265]. More details are provided in [265].

3.2.2.2 Extending to multi-scale search

To ensure the scalability of this strategy to high-resolution (incomplete) volumetric scans, we propose to model the search as a multi-scale navigation process over a discrete scale-space of a given image $\mathbf{I}$. In this context, we redefine the states and actions of the Markov Decision Process $\mathcal{M}$ as follows: s_t encodes the local image context around the current agent position as an axis-aligned box of image intensities. For a given scale level m, $0 \leq m < M$, the image is represented as an instance $L_d(m)$ of an

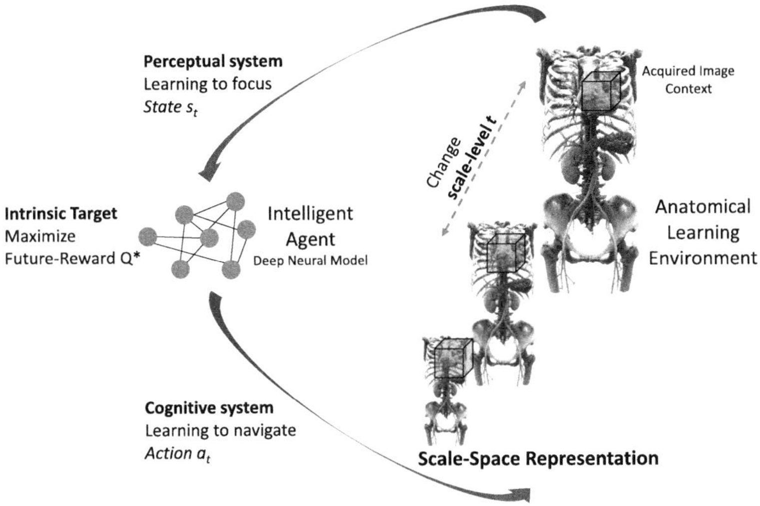

Figure 3.5. Schematic overview of the multi-scale image navigation paradigm based on multi-scale deep reinforcement learning.

M-level scale-space representation L_d of $\mathbf{I}$, with $L_d(0) = \mathbf{I}$ [269]; the action $a_t \in \mathcal{A}$ is the action performed by the agent at time t to move from a voxel position $\mathbf{p}_t$ to an adjacent voxel position $\mathbf{p}_{t+1}$ in image space at the given scale level m. The change in scale from any level m to $m-1$ is facilitated through an *implicit* action – triggered after navigation convergence at level m (see Fig. 3.5).

3.2.2.3 Learning multi-scale navigation strategies

As demonstrated in [262], one can effectively train M separate navigation models corresponding to each of the scale levels. Considering an arbitrary landmark k, the corresponding multi-scale navigation model is defined as: $\boldsymbol{\Theta}_k = [\theta_{k,0}, \theta_{k,1}, \ldots, \theta_{k,M-1}]$, $0 \leq k < P$ with P representing the total number of considered landmarks. We define the search-process as follows: the starting point is in the center of the image at the coarsest scale-level $M-1$. Upon convergence, the scale-level is changed to $M-2$. The search process is continued at level $M-2$. At coarse scale $M-1$, the learning environment covers the entire image. On subsequent scale levels, the exploration is bounded to a local image region around the structure of interest.

Using ϵ-greedy exploration [261], training trajectories are sampled from the learning environment and are stored for each landmark k and scale level m in a cyclic memory array $\boldsymbol{\Xi}(k, m)$. During training, a scale-dependent Bellman cost is optimized [270] using uniform batch-wise sampling of $\Xi(k, m)$:

$$\hat{\theta}_{k,m}^{(i)} = \underset{\theta_{k,m}^{(i)}}{\operatorname{argmin}} \, \mathbb{E}_{(s,a,r,s') \sim \Xi(k,m)} \left[\left(y - Q(s, a; \theta_{k,m}^{(i)} \mid L_d, m) \right)^2 \right], \quad (3.6)$$

where $\theta_{k,m}^{(i)}$ denotes the parameters of the deep neural network search model for landmark k on scale level m at training iteration $i > 0$ and $\Xi(k, m)$ is the associated memory array of past state transitions. The reference value y represents the maximum expected reward for a trajectory starting at the current state, estimated using the update-delay technique [261], based on model parameters $\bar{\theta}_{k,m}^{(i)} := \theta_{k,m}^{(i')}$ from a past training iteration $i' < i$:

$$y = r + \gamma \max_{a'} Q\left(s', a'; \bar{\theta}_{k,m}^{(i)} \mid L_d, m\right). \quad (3.7)$$

We apply the same convergence criterion as defined in [262]. This is determined as the center of gravity of an oscillation cycle.

3.2.2.4 Robust spatially-coherent landmark detection

To better cope with incomplete data, i.e., partial fields of view, we propose to model the spatial distribution of the anatomical landmarks using robust statistical shape modeling. In other words, we constrain the output of the global search model θ_{M-1} to ensure a consistent distribution of the agent positions. Considering a set of N anatomical landmarks in translation and scale-normalized space, we model the distribution of each individual landmark $i \in [0, \ldots, N-1]$ via a multi-variate normal distribution $\mathbf{p}_i \sim \mathcal{N}(\boldsymbol{\mu}_i, \boldsymbol{\Sigma}_i)$, where $\boldsymbol{\mu}_i$ and $\boldsymbol{\Sigma}_i$ are estimated using maximum likelihood. This defines a mean shape-model for the landmark set as $\boldsymbol{\mu} = \left[\boldsymbol{\mu}_0, \ldots, \boldsymbol{\mu}_{N-1}\right]^\top$. Given an unseen configuration of detected points at scale $M-1$ as $\tilde{\mathbf{P}} = [\tilde{\mathbf{p}}_0, \tilde{\mathbf{p}}_1, \ldots]^\top$, a robust shape-model is fitted using M-estimator sample consensus [271] based on 3-point samples from the set of all triples $I_3(\tilde{\mathbf{P}})$. The optimal mean-model fit with maximum consensus is obtained by minimizing the following cost function based on the redescending M-estimator [271]:

$$\hat{\mathcal{S}} \leftarrow \underset{\mathcal{S} \in I_3(\tilde{\mathbf{P}})}{\operatorname{argmin}} \sum_{i=0}^{|\tilde{\mathbf{P}}|} \min \left[\frac{1}{Z_i} \left(\phi(\tilde{\mathbf{p}}_i) - \boldsymbol{\mu}_i\right)^\top \boldsymbol{\Sigma}_i^{-1} \left(\phi(\tilde{\mathbf{p}}_i) - \boldsymbol{\mu}_i\right), 1 \right], \quad (3.8)$$

where $\phi(\mathbf{x}) = \frac{\mathbf{x}-\mathbf{t}}{s}$ is a projector to normalized shape-space with the estimated fit $\hat{\mathbf{w}} = [\mathbf{t}, s]$ on set $\mathcal{S}$. More details on this model of the spatial consistency can be found in [264].

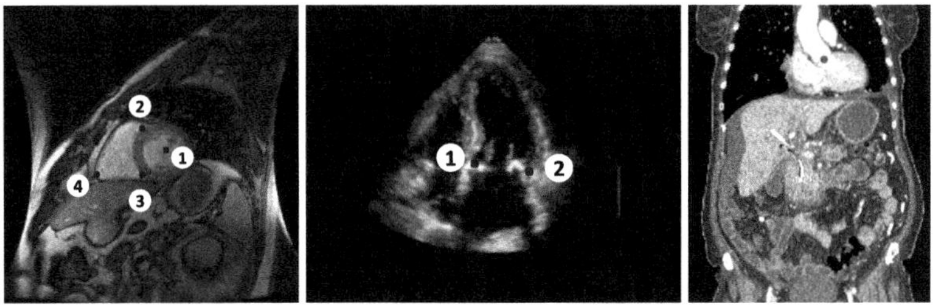

Figure 3.6. Left: The LV-center (1), the anterior/posterior RV-insertion points (2)/(3) and the RV-extreme point (4) in a short-axis cardiac MR image. Middle: The mitral septal annulus (1) and the mitral lateral annulus points (2) in a cardiac ultrasound image. Right: The center of the aortic root in a frontal slice of a 3D-CT scan.

3.2.2.5 Experiments

We evaluated the system on 2D image data based on several cardiac landmarks visible in short axis-view MR images or apical four-chamber view cardiac ultrasound images (see Fig. 3.6 for complete definition). The MR dataset includes 891 images from 338 patients, while the US dataset covers 1186 images from 361 patients. Based on a random patient-wise split, the method could detect the 4 landmarks from 2D MR images with an average error of 3.0 mm. This represents an error reduction of over 50% compared to the state-of-the-art [272,273]. A similar performance level was achieved on the 2D US data. More details are provided in [265].

We also measured the performance on 3D image data, using a dataset of 5043 3D-CT volumes from over 2000 patients. The dataset contains a wide variety of scan types with different fields of view, such as cardiac CT scans (with and without contrast), thoracic scans, abdominal scans, CT scans of the legs and pelvis, CT scans of the head and neck and whole body scans. Based on the available native resolution of each scan, we generated a scale-space that includes up to 6 levels of scale at isotropic resolutions of 0.5 mm, 1 mm, 2 mm, 4 mm, 8 mm and 16 mm. The voxel values were clipped to the 0–800 HU interval and then normalized to unit interval [0, 1]. The evaluation is performed on a set of 29 landmarks located on heart structures or vessels throughout the body. Fig. 3.7 gives an overview. The number displayed in subscript after each landmark indicates the highest scale level used in the detection process. More details on the definition of these landmarks are covered in [263].

A random split at patient-level was used for validation, resulting in 70% training and 30% test data. We refer to false positive (FPR) and false negative rates (FNR) to measure the accuracy

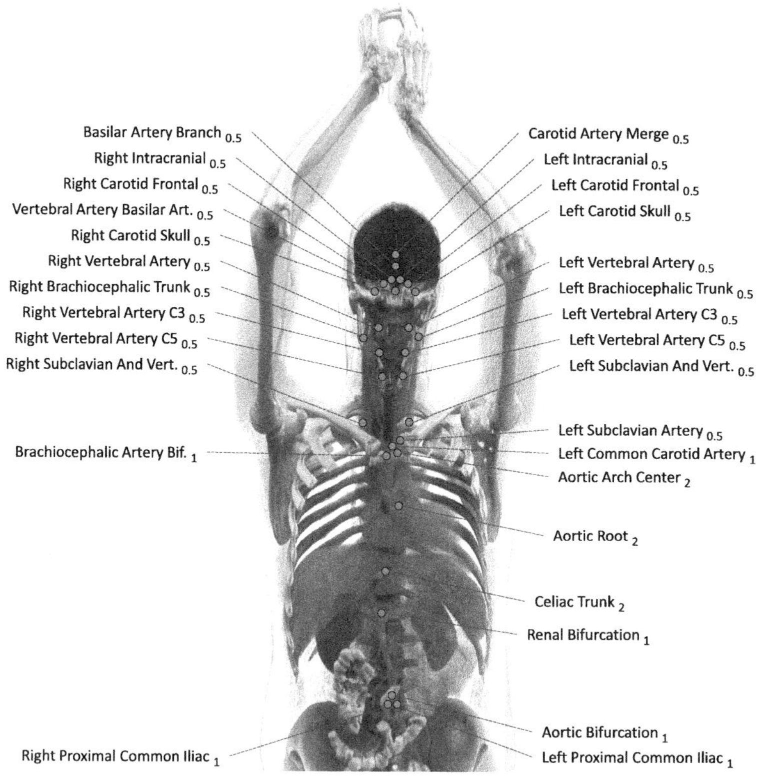

Figure 3.7. Visualization of the considered cardiac and vascular landmarks.

in recognizing the absence of landmarks from the field of view. Please note that for these two measures we eliminated landmarks that were closer than 3 cm to the scan border. The elimination of these cases is motivated by the poor annotation quality around the border due to occlusion. For details on the architecture of the convolutional neural network used to approximate the optimal Q function and the model parameters we refer to [262,263].

The navigation starts at runtime on coarse scale in the center of the scan. Let $\tilde{\mathbf{P}}$ define the landmark locations after convergence on coarse scale level. A robust statistical model is fitted to this point set, allowing the robust detection and correction of outliers (please note that for this step a larger set of landmarks is used; that set is specified in [263]). Subsequently, the navigation is continued across scale levels under spatial coherence. The results are shown in Table 3.3, highlighting the high detection accuracy and recognition rate on the presence of landmarks in the field of view (excluding border cases).

Table 3.3 Accuracy at detecting different anatomical landmarks in incomplete 3D-CT scans. The accuracy is measured in mm.

Landmark	FPR	FNR	Mean	STD	Med.
Carotid Artery Merge	**0%**	**0%**	2.68	2.96	1.65
Basilar Artery Branch	**0%**	**0%**	1.75	2.17	1.16
Right Intracranial	**0%**	**0%**	1.88	1.65	1.54
Left Intracranial	**0%**	**0%**	2.52	1.90	2.31
Right Carotid Frontal	**0%**	**0%**	1.60	1.63	1.11
Left Carotid Frontal	**0%**	**0%**	1.07	0.99	0.69
Right Carotid Skull	**0%**	**0%**	1.44	1.51	1.07
Left Carotid Skull	**0%**	**0%**	1.03	1.01	0.74
Right Vertebral Artery	**0%**	**0%**	0.87	1.32	0.60
Left Vertebral Artery	**0%**	**0%**	1.24	0.95	1.10
Right Brachiocephalic Tr.	**0%**	**0%**	1.89	2.67	0.66
Left Brachiocephalic Tr.	**0%**	**0%**	1.92	2.77	0.67
Right Vertebral Artery C3	**0%**	**0%**	1.00	1.59	0.75
Left Vertebral Artery C3	**0%**	**0%**	1.00	1.26	0.67
Right Vertebral Artery C5	**0%**	**0%**	3.30	4.90	1.42
Left Vertebral Artery C5	**0%**	**0%**	1.98	3.90	0.77
Vertebral Art.–Basilar Art.	**0%**	**0%**	2.24	2.89	1.06
Right Subclavian–Vertebralis	**0%**	**0%**	4.68	4.39	2.93
Left Subclavian–Vertebralis	**0%**	**0%**	3.99	3.85	2.68
Left Common Carotid Art. Bif.	**0%**	**0%**	2.99	2.51	2.20
Brachiocephalic Art. Bif.	**0%**	**0%**	2.25	1.99	1.57
Left Subclavian Art. Bif.	**0%**	**0%**	2.55	2.76	1.50
Aortic Arch Center	**0%**	**0%**	1.96	1.22	1.72
Aortic Root	**0%**	**0%**	3.34	2.21	2.97
Celiac Trunk	**0%**	**0%**	2.36	1.88	1.97
Aortic Bifurcation	**0%**	**0%**	1.33	1.27	1.07
R. Proximal Common Iliac	**0%**	**0%**	2.25	2.54	1.27
L. Proximal Common Iliac	**0%**	**0%**	2.40	2.70	1.51
Renal Bifurcation	**0%**	**0%**	1.81	2.31	1.29

In terms of runtime, the reformulation of the detection as a multi-scale navigation process enables an average detection time per landmark of 52 milliseconds on CPU (Intel 8-core) and **28 milliseconds** on GPU (Nvidia Pascal).

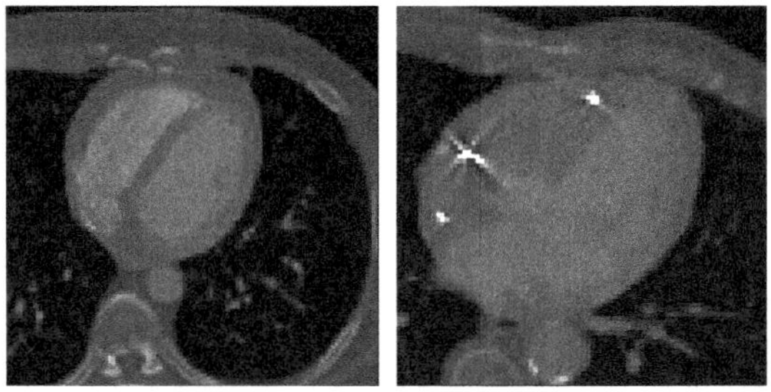

Figure 3.8. Segmentation masks for heart isolation computed with a deep neural network.

3.2.3 Deep image-to-image segmentation

An alternative approach for image segmentation is based on fully-convolutional deep neural networks (FCNs) [274]. In this context, the segmentation problem is formulated as an end-to-end functional mapping from image pixels to an image segmentation mask via an FCN architecture. Typically, the architecture is composed of an encoder part which processes the input image signal to a latent representation (also called embedding), and a decoder part which learns to map this embedding to a segmentation map over the anatomical structures of interest. One may use various cost functions for optimization, e.g., per-pixel mean squared error or the Dice coefficient. Several architectural improvements have been proposed to optimize the gradient flow and allow for more effective learning, such as skip connections [275] or densely connected blocks [276]. On volumetric medical image data, however, the training of these architectures becomes a tedious operation due to very high memory requirements of 3D spatial processing. To address this problem, several solutions have been proposed. Dormer et al. [277] propose to train the network on image sub-regions (patches), while Zheng et al. [278] present a robust aggregation scheme for 2D segmentation masks that have been computed sequentially on 2D image slices. An elegant and memory-efficient solution based on a patch-wise approach is presented in [279]. There, a deep memory network is employed to ensure that in a patch-wise prediction paradigm also essential global image/shape features are captured. This is an important prerequisite to ensure robustness and generalization.

Based on previous work [280] proposed in the context of liver segmentation, a deep FCN model can be designed for heart segmentation/isolation. The network is fully 3D, and follows the typical encoder-decoder structure. The architecture also includes skip connections which improve the gradient flow and allow for faster and more effective learning. A voxel-wise cross entropy loss is used to drive the training. Experiments conducted on a 3D-CT dataset of 896 cases, using 796 cases for training (10% of these for validation) and the remaining 100 cases for testing demonstrate competitive advantages with an average Dice score of 0.93. Fig. 3.8 shows two slices of one test case, with the heart isolation mask as a transparent overlay.

3.3 Structure tracking

Structure tracking is a fundamental subject in cardiac imaging research. Efficient and robust tracking on the motion and deformation of the heart serves a large variety of clinical applications including myocardial strain measurement, anomalous state detection, infarction prediction, real-time image-guided intervention and cardiovascular surgery planning. Recent advances in imaging technologies allow cardiologists to capture morphological and functional information of complex structures and their dynamic changes effectively. For example, echocardiography captures the cardiac motion in real time and provides important guidance during surgeries for valve repairments. X-ray angiography serves as the primary modality in percutaneous coronary interventions (PCI) and catheter-based electrical physiology (EP) therapies to precisely visualize and target the surgical object [281]. While these imaging techniques have achieved great success in clinical practice, they create challenges for the research community to develop models to extract and process important structural as well as temporal information from the captured dynamic images. Such challenges include sophisticated image conditions (clutters, illumination and noise), complex anatomy motion and deformation, partial/full object occlusions, and real-time processing requirements.

In computer vision literature, object visual tracking has been extensively studied. Many different approaches have been proposed in the past decades, see [282–285] for detailed reviews. Tracking methods can be classified into three categories based on the representations of the objects: point tracking, kernel-based tracking and silhouette tracking models. Point tracking models aim to find correspondences on the detected key points of the target object at each frame. Examples of key points include object centroid, critical landmarks and point cloud on object boundaries. Kernel-based tracking models mask the target object with carefully designed kernels and search for representation matchings with efficient methods such as mean-shift. Silhouette tracking is closely related to kernel tracking. It performs shape or contour matchings of target object cross frames. A substantial limitation of all these tracking methods comes from the fact that the kernels or representations are 'engineered' and may not capture enough deep insights of the images.

In recent years, significant attention has been focused on the development of deep learning based tracking models. Compared with conventional methods, deep neural network models can extract more informative features and have shown superior performance in various applications. Based on the network topology and methodology types, we organize them into three different categories.

Tracking with convolutional Neural Networks (CNNs). To address the aforementioned limitations from the 'engineered' representations of target objects, leveraging high level features from CNNs serves as a natural remedy. Siamese network [286] is one of the most commonly used architectures for similarity-based tracking. It processes two different inputs through the same network computations and provides a similarity score based on the extracted features. One of the early work is due to Bertinetto et al. [287], who propose a fully convolutional Siamese network to find a target object in consecutive frames with a region-wise similarity measure. Similar strategies have been widely developed, including GOTURN tracker with box regression on targets [288], DSiam tracker with online Siamease network updating [289], variants of CFNET with add-on correlation filters [290,291] and different variants of SiamRPN with region proposals after feature extraction [292–294]. Besides similarity learning based Siamese networks, different models considering domain and appearance changes have been studied. Nam et al. [295] proposed MDNet which learns a domain independent representation which encodes the moving object and uses it for detection in the next frames. CREST [296] represents the discriminative correlation filter (DCF) [297, 298] as convolution and applies residual learning to accommodate appearance changes. Zhu et al. [299] also take optical flow information into account and proposed a model on correlation tracking with spatial-temporal attention. The application of such models in cardiac imaging is under development. Recently, Parajuli et al. [300] applied a flow network on left ventricle motion analysis where the motion is modeled as flow through graphs and the similarities between graph nodes are learned from a Siamese network.

Tracking with Recurrent Neural Networks. It is found that recurrent neural networks can well encode temporal state information and thus are effective for sequential data. Cui et al. [301] proposed a Recurrently Target-attending Tracking (RTT) model. It estimates a confidence map for object motion using a multidirectional RNN and uses it to improve correlation filters for detection cross frames. Ondrúška et al. [302] leveraged RNNs as a feature encoder to reveal object occlusion by learning from raw sensor data. Kahou et al. [303] and Gan et al. [304] used RNNs and attention schemes to predict the distribution of the object's locations. Binary classifiers are constructed to find the final location. Similarly, Ning et al. [305] proposed a spatially supervised recurrent convolutional neural network to exploit the history of object locations for object tracking.

Tracking with Reinforcement Learning. Besides landmark detection and image registration, deep reinforcement learning has also been applied in visual tracking. Zhang et al. [306] proposed to learn tracking policies with a detection-memorization CNN-LSTM scheme. The reward is calculated by the detection error at each frame. Similar to the similarity learning models with Siamese networks, Choi et al. [307] introduced a reinforcement learning scheme for tracking based on a template selection strategy. After extracting features of candidate regions and template via a Siamese network, a policy network is attached to determine the similarities between them. Instead of taking the output directly from the last layer of the policy network for decision, Huang et al. [308] further improved the robustness and speed up achieved by such model scheme by adding an agent to choose the decision layer selectively. Yun [309] proposed an Action-Decision Network to use pre-trained networks to control actions of the agents and add online fine-tuning during tracking.

Other methods such as correlation filtering [297,310–312] and the more recent adversarial learning [313] have also demonstrated great value in object tracking. While these models have been successfully applied in object tracking in natural scenes, how to effectively adapt these models in the medical domain, especially for cardiac activity tracking, is still an open question.

3.4 Summary

This chapter presented several methods for cardiac anatomy detection in medical images. This includes classification-based solutions, such as the marginal space (deep) learning frameworks, and a reinforcement learning-based solution referred to as intelligent multi-scale image navigation. Through extensive experiments, we demonstrated the power of deep learning architectures in capturing the image information and performing prediction. We also analyzed the significant advantages of approaching the detection problem as a reinforcement learning task, highlighting the improved precision, robustness and in particular detection speed. Using robust statistical shape modeling within the multi-scale image navigation paradigm, a system can be defined, that can accurately recognize whether any anatomical structures of interest are missing from the field of view.

For the task of image segmentation, we show the performance of the marginal space (deep) learning frameworks and of modern image-to-image deep learning models in segmenting anatomical structures like the heart chambers or valves. Here we discuss the advantage of image-to-image solutions in implicitly capturing prior shape information and use it for a more robust prediction.

In the space of image tracking, we highlight in this chapter how modern approaches in artificial intelligence have demonstrated their abilities in encoding complex visual object features and decoding their motion patterns both efficiently and effectively. They have shown great potential towards real-time modeling for anatomy motion and deformation, and thus offer support for accurate cardiac function assessment, anomaly detection and prediction.

4

Data-driven reduction of cardiac models

Lucian Mihai Itu[a], Felix Meister[a,b], Puneet Sharma[a], Tiziano Passerini[a]

[a]*Siemens Healthineers, Princeton, NJ, United States.* [b]*Friedrich-Alexander University Erlangen-Nuremberg, Erlangen, Germany*

Physiological models of cardiac biology and biomechanics allow high fidelity representation and simulation of the mechanisms governing the functions of the heart at the cellular, tissue, organ and system level. The underlying mathematical formulation of the models typically requires complex numerical approximation strategies and high computational costs, as discussed in the first part of this book. This often induces a trade-off between the descriptive power of the model and its utility. Simplifying assumptions can reduce the complexity of the numerical approximation, and the time required to compute a solution; at the expense of the capability of the model to describe the complex, multi-scale dynamics typical of biological systems, and to capture inter-patient variability. Different approaches are described in this chapter to reduce the computational complexity of physiological models, while still providing accurate and detailed representation of the physical phenomena of interest. These approaches are described through three use cases, covering cardiac hemodynamics, electrophysiology and biomechanics.

Coronary artery disease (CAD) is the most prevalent cardiovascular disease (CVD). Plaque builds up in the coronary arteries and limits the flow to the myocardium, especially when the demand is increased (exercise, stress), potentially leading to myocardial infarction, or even death. For the non-invasive evaluation of the functional significance of coronary artery disease, presented in section 4.1, the explicit computation of patient-specific hemodynamics in the coronary tree is replaced altogether with a model trained using deep learning, providing statistically indistinguishable results, and enabling drastic reduction in the compute time. The deep learning model expresses the relation between anatomical features of the coronary tree that can be automatically extracted from medical images, and the patient-specific blood pres-

Artificial Intelligence for Computational Modeling of the Heart
https://doi.org/10.1016/B978-0-12-817594-1.00015-2

sure values as produced by state-of-the-art physiological models of blood flow. To enhance its performance and generalization properties, the deep learning model is trained on an extensive database of synthetically generated anatomical models of the human coronary tree, spanning the geometric variability of a reference human population.

Atrial fibrillation is an increasing socio-economic burden. Patient-specific modeling of atrial electrophysiology has the potential to help understanding the disease but also devise effective treatments. Unfortunately, these models are often computationally demanding, controlled by a large number of parameters, and therefore not directly applicable within a clinical workflow. Section 4.2 describes the definition of a surrogate model of cellular electrical activity to speed up atrial electrophysiology simulation. The time pattern of the cellular action potential as expressed by state-of-the-art computational models is estimated using regression from the model parameters. The estimated action potential is then used in combination with standard computational models of organ-level electrophysiology to replace the explicit computation of the cellular electrical potential dynamics, enabling drastic speed-up while accurately reproducing key features of the action potential morphology.

Finally, the prediction of heart motion and its mechanical characterization can provide valuable information on the state and function of the organ, as well as insights on the effect of different therapy options for the specific patient. As detailed in section 4.3, a model can be trained using deep learning to predict myocardium acceleration from the kinematic and dynamic state of the system. The predicted acceleration is used in an explicit time advancing scheme for the solution of the elastodynamics of structures with mechanical properties compatible with those of biological tissue. Compared with a state-of-the-art numerical solver for the computation of soft tissue deformations, this approach allows the use of time steps one order of magnitude larger, relaxing the stability condition of the numerical scheme and potentially allowing significant speed-up.

4.1 Deep-learning model for real-time, non-invasive fractional flow reserve

4.1.1 Introduction

Hemodynamically significant coronary lesions are treated typically through Percutaneous Coronary Intervention (PCI). Clinical decision making is often based on anatomical features, like the percentage diameter stenosis, determined during invasive coronary angiography (ICA). The degree of reduction in lumen diameter may be determined either visually or through computer-assisted quantitative coronary angiography [314].

Non-invasive imaging techniques, like Coronary Computed Tomography Angiography (CCTA), play nowadays an increasingly important role for CAD diagnosis, prior to ICA. CCTA has been shown to provide a high negative predictive value, but, by overestimating the lesion severity [315,316], it also leads to a large number false positives.

The purely anatomical CAD assessment, irrespective of the medical imaging modality, does not correlate closely with the functional significance of the coronary lesions. Hence, the diagnostic measure of Fractional Flow Reserve (FFR) has been proposed as an alternative [317]. FFR is defined as the ratio of cycle-averaged pressure distal to the stenosis and cycle-averaged aortic pressure, both measured during hyperemia. It represents a surrogate measure for the reduction of hyperemic flow caused by the stenosis, and has been shown to lead to superior long-term patient outcomes, when compared to the purely anatomical assessment [318]. Although FFR is now the gold standard in international guidelines [319,320], it is still not used at a large scale, mainly due to the higher costs and risks, and the longer procedure times [321].

In view of these limitations, approaches based on computational fluid dynamics (CFD) have been introduced for determining FFR non-invasively from medical images acquired at rest [322–329]. These models are run under patient-specific conditions: the coronary anatomical model is reconstructed from medical images, and physiological information is derived and employed for model personalization. The CFD models consist of partial differential equations, which can be solved only numerically, leading to a very large number of algebraic equations. Thus, the solution requires several hours in case of high-fidelity three-dimensional models, and several minutes in case of reduced-order models [315,330]. Due to the computationally intensive aspect of CFD models, and the time-consuming process required for generating the anatomical model, they are not used for intra-operative assessment and planning, where real-time performance is required. Alternatively, artificial intelligence based solutions may be employed, capable of providing results instantaneously. To develop such solutions, a large training database is required, containing input–output data pairs, where input data is represented by the vascular geometry, and the output is FFR [331]. Once the training phase has been completed, the online application provides results in real-time. Such supervised machine learning (ML) algorithms are routinely employed in medical imaging applications, e.g. organ segmentation [332]. Moreover, machine learning models can also be employed to reproduce the behavior of nonlinear computational models [37,333].

Herein, a machine learning model for near real-time FFR prediction is presented [334]. The main goal was to obtain a method providing results which are statistically indiscernible from those obtained with the CFD based approach. To build a large enough training database synthetically generated pathological coronary anatomical models are employed. Cycle-averaged pressures, and derived FFR, are computed at each centerline location of each synthetic tree using a reduced-order CFD model [315]. Anatomical features are then determined at each location of each coronary tree, and paired with the corresponding CFD-based FFR values to build the training database. Finally, a deep neural network (DNN) is trained to predict FFR. The resulting model computed FFR in 2.4 ± 0.44 seconds, whereas the CFD model required 196.3 ± 78.5 seconds, both on a standard desktop computer with a 3.4 GHz Intel i7 8-core processor. Ideally, the machine learning model should be trained on CFD results obtained for patient-specific anatomical models. Since the costs and the effort for setting up such a large training database are prohibitive, purely synthetically generated anatomical models of the coronary tree are used during the training phase, and generated the corresponding ground truth FFR values using the CFD model. The generation of the synthetic database follows a parameterization approach relying on anatomical features of the arterial coronary tree, which enables the generation of various configurations like serial stenoses, three-vessel disease, bifurcation stenoses, diffuse stenoses or rare pathologies.

In the following, $cFFR_{CFD}$ refers to the FFR values computed with the CFD based model, while $cFFR_{ML}$ refers to the FFR values computed with the ML based model. Verification and validation are performed in three steps: (i) $cFFR_{ML}$ vs. $cFFR_{CFD}$ on synthetic anatomical models, (ii) $cFFR_{ML}$ vs. $cFFR_{CFD}$ on patient-specific anatomical models, (iii) $cFFR_{ML}$ vs. invasive FFR on patient-specific anatomical models.

4.1.2 Methods

In the following a framework employed for machine learning based FFR computation in coronary arteries is described:

- Generation of synthetic coronary arterial trees
- CFD based approach for computing FFR in coronary arteries
- Feature definition and extraction for generating the mapping between the coronary anatomy and FFR

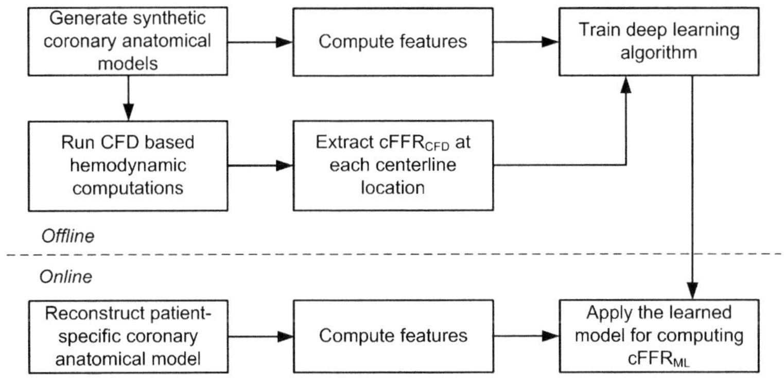

Figure 4.1. Overall workflow of the proposed method.

While the training process is based solely on synthetic arterial trees, FFR is computed during the online prediction phase for patient-specific coronary anatomies. This step is fully automated, consisting of feature extraction, application of the pre-learned model to compute $cFFR_{ML}$ at all locations of the coronary tree, and visualization of the color coded coronary tree (see Fig. 4.1).

The generation of the patient-specific coronary geometry is semi-automatic: an initial version of the coronary centerlines and luminal contours is computed automatically, and then edited by the user after careful inspection [335]. Once the coronary geometry is available, features are extracted.

4.1.2.1 Generating synthetic coronary arterial trees

The synthetic coronary arterial trees used for setting up the training database are generated algorithmically. As displayed in Fig. 4.2 this is done in three sequential stages. During the first stage, the structure of the coronary tree is determined, i.e. number of generations and number of segments. During the second stage, first the length of each segment is set, and, next, the vessel radius at each location is defined (including tapering). These properties are determined by a set of parameters, whose values are randomly sampled in a pre-defined interval (Table 4.1): the interval limits have been chosen to ensure that a large range of anatomical variations is covered, leading to a large range of variations in the derived hemodynamic quantities.

The first two stages enable the generation of healthy coronary anatomical models. The third stage inserts stenoses into the coronary trees. Thus, the number of stenoses is set randomly between zero and three for LAD, LCx and RCA, and between zero and two for their side branches. Each stenosis is defined by a set of pa-

Table 4.1 Parameters with pre-defined value ranges employed in the generation of the synthetic coronary trees.

Step	Parameter	Range
	Number of main branches	3 (LAD, LCx, RCA)
Step 1	Number of side branches (1st gen.)	2–5
	Number of side branches (2nd gen)	0–2
	Root radius [336]	0.15–0.35 cm
	Power coefficient [332,337,338]	2.1–2.7
Step 2	Area ratio [339]	0.35–0.45 (main branch) 0.6–0.8 (side branch)
	Degree of tapering [340]	−20% to +5% from top to bottom
	Length [341]	1.5–4 cm
	Bifurcation angle [336,342]	30–90 degrees

rameters: percentage diameter stenosis, stenosis length, stenosis center, length of the stenosis region with minimum radius, proximal – distal radius variation (radius tapering). Both single branch and bifurcation stenoses are generated. The Medina classification is employed in case of bifurcation stenoses, and for each stenosed bifurcation segment the above mentioned parameters are independently set. Fig. 4.2 displays an overview of the stenosis specific properties. Parameter sampling is based on a uniform distribution, except for the left main and RCA ostial radius, and the percentage diameter stenosis, for which a normal distribution has been chosen.

A database of 12.000 synthetic coronary arteries was generated reflecting anatomical variations of stable CAD patients. While a wide range of pathological configurations encountered in clinical practice is covered, some seldomly encountered coronary anatomies, e.g. aneurysms, are not covered.

4.1.2.2 CFD-based hemodynamic computations

The methodology described in the previous paragraph covers the generation of the input data in the training database. To generate the corresponding output, a reduced-order multiscale fluid-structure interaction hemodynamic model is employed to compute the CFD based FFR values. The model was described in details in [315], and then validated in several clinical studies by comparison against invasively measured FFR. The diagnostic ac-

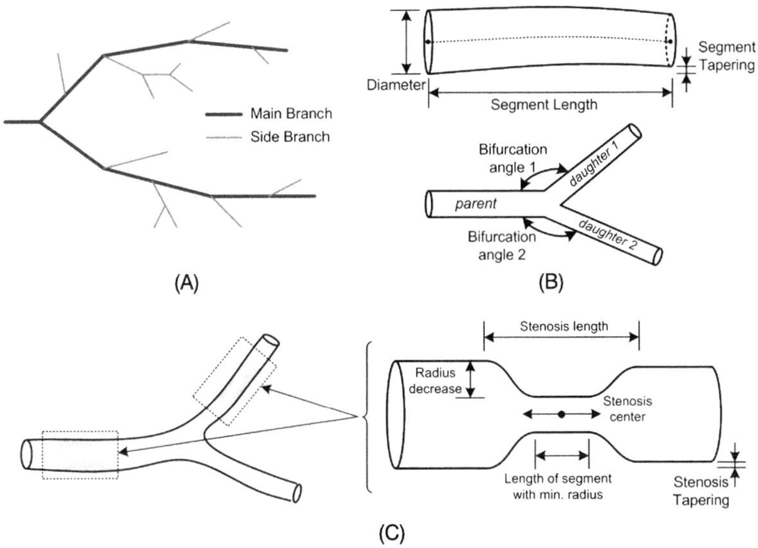

Figure 4.2. Three stage approach for generating synthetic coronary geometries: (A) Define coronary tree skeleton, (B) Define healthy coronary anatomy, (C) Define stenoses.

curacy of non-invasively computed FFR varied between 75% and 85% [322,328,343–347].

The hemodynamic model displayed in Fig. 4.3 is represented by a set of partial differential equations which can be solved only numerically. The reduced-order Navier–Stokes equations employed herein enable the computation of time-varying pressures, flow rates and cross-sectional areas. A population-average viscosity value is employed and a lumped parameter model representative for the coronary microcirculation is coupled at the outlets of the epicardial coronary arteries [348]. The reduced-order Navier–Stokes equations are valid as long as no abrupt radius variations are present. To ensure that pressures are computed accurately in the stenosis regions, the momentum conservation equation is modified to enable a correct computation of the additional energy losses caused by the flow turbulence. Thus, the complex shapes of stenoses are taken into account appropriately and the pressure loss across the stenoses are predicted correctly. The coronary tree is coupled to a simple, population-average systemic circulation model composed from the aorta and the distal circulation. The inlet boundary condition of the aorta is set by a lumped parameter model of the heart: the ventricular pressure is applied as intramyocardial pressure in the lumped parameter model of the coronary

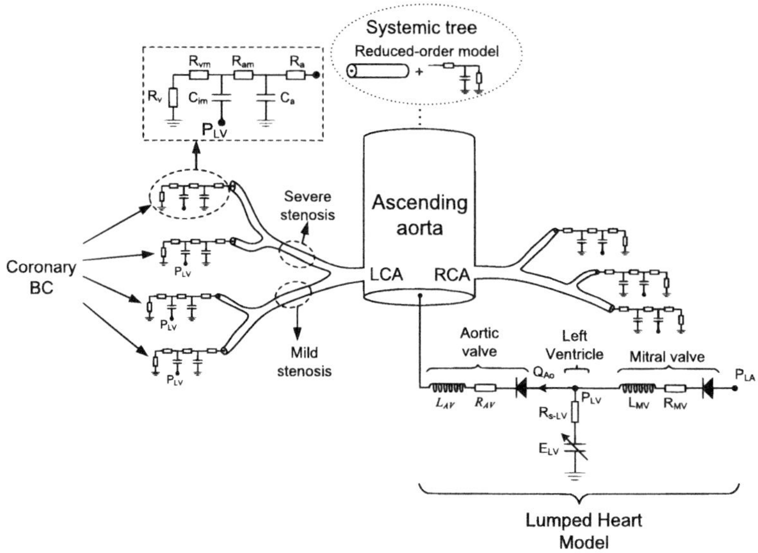

Figure 4.3. Multiscale model of the systemic and coronary arterial circulation.

microcirculation to correctly capture the effect of myocardial contractions on the coronary flow conditions.

Since invasive coronary measurements are not available and the goal was to develop a fully non-invasive approach, the personalization of the boundary conditions is performed based on allometric scaling laws. Thus, first a healthy reference radius is estimated for each branch. Next, the total coronary flow at rest is defined based on the reference radiuses of all branches [337,349], and is then distributed to all outlets of the coronary anatomical model based on the Murray law [338]. Finally, the microvascular resistances at each outlet are determined through an iterative calibration procedure which automatically tunes the parameters [350]. Since FFR is measured invasively during hyperemia, this state is also simulated in the CFD model, by appropriately decreasing the total microvascular resistance for each outlet boundary condition [343]. CFD based FFR is then determined at each centerline location as ratio of mean pressure at that location and the mean pressure in the aorta.

4.1.2.3 Machine-learning based FFR computation

Although in practice FFR is determined at a limited number of locations, to allow the radiologist to probe any coronary location, the goal was to predict FFR independently at any centerline location in the reconstructed anatomical model. Thus, a set of features

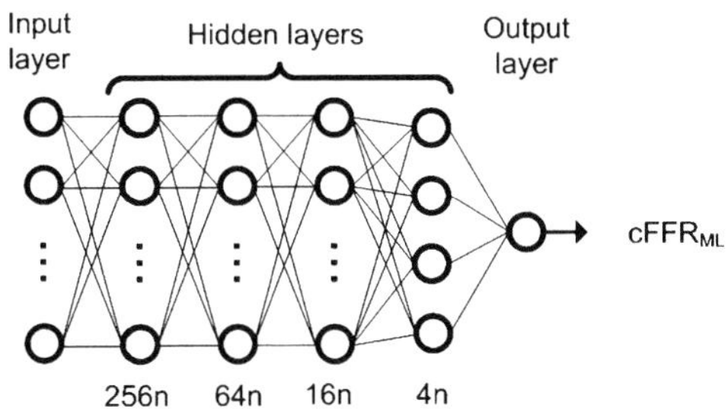

Figure 4.4. Deep neural network model employed for computing $cFFR_{ML}$: fully connected architecture with four hidden layers.

is defined independently at each centerline location. Since local coronary hemodynamics are influenced by both the local and the proximal and distal anatomy, features are defined based on local, proximal and distal anatomical characteristics. The coronary circulation has a tree like structure, and, thus, there is a single upstream path, but typically multiple downstream paths. Hence, to define the features of the distal anatomy, a main downstream path is defined, determined based on the healthy reference radius, the number and the length of downstream branches.

A deep neural network containing 4 hidden layers is used as machine learning model (Fig. 4.4). Each neuron in each layer is connected to all neurons in the next layer, i.e. a fully connected architecture is employed (no convolutional layers). A total of 28 features is extracted from the anatomical model for each location, and connected to the input layer of the network. The first hidden layer contains 256 neurons, and, then, the number of neurons is decreased by a factor of 4 in each subsequent layer. All activation functions are of type sigmoidal, and the output layer, represented by a single neuron, has a linear activation function. Random (Xavier) initialization is performed for all weights. By first training each layer as an autoencoder, the overall model training time is reduced. The loss is defined as the mean squared error between the CFD based FFR values and the ML predicted values, and the parameter optimization is performed using a Stochastic Gradient Descent algorithm. To further reduce the model training time, an optimized GPU based implementation was employed. The synthetic datasets were randomly split into training / validation datasets using a 5:1 split. During training and validation, when learning rate, momentum and other relevant hyperparameters were tuned, the model was never evaluated on patient specific datasets.

In the following the specific features that were defined as input to the network are described in detail.

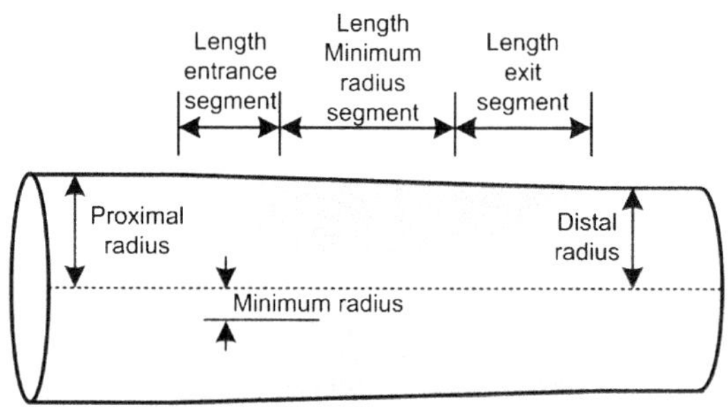

Figure 4.5. Features describing a stenosis.

4.1.2.4 Local features

These features characterize the anatomy only at the specific location at which the FFR prediction is performed. Herein, the actual radius of the vessel, and the reference radius of the hypothetically healthy vessel are used (if the current location is not stenosed, the two values are identical). Additionally, a branch specific ischemic weight was defined, representing the potential contribution of the branch to the overall ischemic state of the individual. Its value is initially set based on the reference radiuses of all branches in the anatomical model, and then adapted as described in the following.

4.1.2.5 Features defined based on the proximal and distal vasculature

Given the significant inter-dependence of coronary hemodynamics at different locations features describing the proximal and distal vasculature are of significant importance. A first step in the definition of the proximal and distal features is the identification of all proximal and distal stenoses. Stenoses are identified automatically: all narrowings with a radius reduction larger than 10% are marked as stenoses. Next, all identified stenoses are ranked based on the degree of radius reduction, and the most severe four proximal and distal stenoses are retained. For each stenosis the following anatomical characteristics and their non-linear product combinations are computed (Fig. 4.5):

- Minimum radius
- Proximal radius
- Distal radius
- Length of the stenotic segment with minimum radius
- Total stenotic length

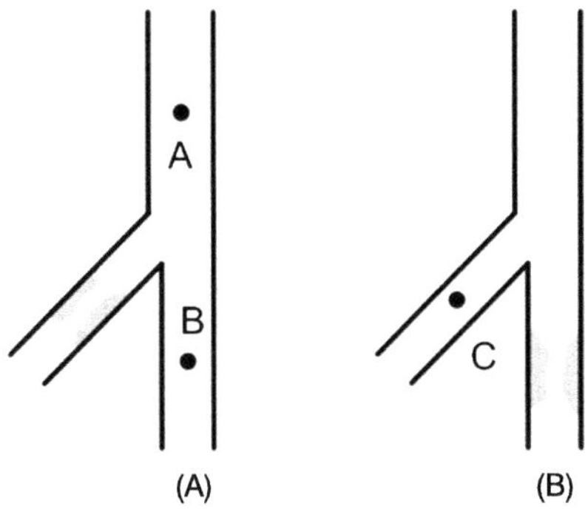

Figure 4.6. Examples of hemodynamic interdependence between coronary branches.

- Radius reduction [%]:

$$\%RR = \left(1 - \frac{r_{sten}}{(r_{prox} + r_{dist})/2}\right) \cdot 100 \tag{4.1}$$

where r_{prox} and r_{dist} are the proximal and respectively distal radiuses of the stenosis, while r_{sten} is the smallest stenotic radius.

Since the threshold is set at 10%, also very mild stenoses are taken into consideration, which, taken separately, have a small ischemic effect, but, when cumulated, may lead to a functionally significant ischemic state. Additionally, cumulative proximal and distal features are also defined, based on the aggregation of the features describe above. Furthermore, the proximal and distal lengths are also defined as input features. As mentioned above, coronary hemodynamics present a significant inter-branch dependence. For the schematic geometry displayed in Fig. 4.6(A), coronary hemodynamics at locations A and B are affected by the side branch stenosis. The stenosis reduces the flow rate in the side branch, and, hence, also in the parent branch. Due to the reduced flow in the parent branch, the pressures at points A and B will be higher. In another example, displayed in Fig. 4.6(B), the stenosis on the main branch affects the flow characteristics at point C: due to the stenosis both the flow and the pressure loss in the upstream branch are lower, and, hence, the overall pressure in the side branch decreases. To account for these characteristics of coronary hemodynamics, the initial ischemic weight values, defined as described above, are adapted automatically. A non-linear adaptation function is defined, taking as input the initial value of the ischemic weight as well as the features of the proximal and distal stenoses and the stenoses lying on side branches.

4.1.3 Results

The proposed methodology was evaluated in three steps, described in the following.

4.1.3.1 Validation on synthetic anatomical models

The synthetic datasets were split randomly using a ratio of 5:1 into training and validation datasets. Hence, 2000 out of the 12 000 synthetically generated anatomical models were used for an initial validation step. CFD based FFR was compared with ML based FFR at all centerline locations and obtained an almost perfect correlation results ($\rho = 0.9998$, $p < 0.001$). The mean difference between $cFFR_{ML}$ and $cFFR_{CFD}$ was 0.0008 (no systematic bias). When considering CFD based FFR as ground truth, and employing 0.8 as binary cut-off value, ML based FFR has an accuracy of 99.7%.

4.1.3.2 Validation on patient specific anatomical models

A dataset based on 87 patients was further employed for validation. An anatomical model was reconstructed for each patient from the CT data, using centerline detection and lumen border segmentation, as described in [322,328]. As for the synthetic evaluation dataset, the correlation between $cFFR_{ML}$ and $cFFR_{CFD}$, at all coronary centerline locations, was excellent ($\rho = 0.9994$, $p < 0.001$) and the Bland–Altman analysis revealed no systematic bias: mean difference -0.00081 ± 0.0039 (Fig. 4.7, Fig. 4.8). The mean computation time was 2.4 ± 0.44 seconds for ML based FFR, and 196.3 ± 78.5 seconds for CFD based FFR, both on a standard desktop computer (3.4 GHz Intel i7 8-core CPU). For a more detailed analysis the individual centerline locations have been grouped into 5 bins, based on the $cFFR_{CFD}$ values: 0.0–0.6; 0.6–0.7; 0.7–0.8; 0.8–0.9; 0.9–1.0. The bins were defined either globally or separately for the three main coronary branches: LAD (Left Anterior Descending), LCx (Left Circumflex) and RCA (Right Coronary Artery). The results of this detailed comparison between $cFFR_{ML}$ and $cFFR_{CFD}$ analysis are displayed in Table 4.2: overall the agreement between $cFFR_{CFD}$ and $cFFR_{ML}$ is very high for all locations and all bins.

4.1.3.3 Validation against invasively measured FFR

FFR was measured invasively for 125 stenoses in the above mentioned dataset containing 87 patients. The exact measurement location was set either around 2 cm distal from the stenosis [317] or, if available, extracted from the coronary angiographies displaying the pressure wire. All lesions with FFR ≤ 0.80 were considered hemodynamically significant [327] (38 out of 125

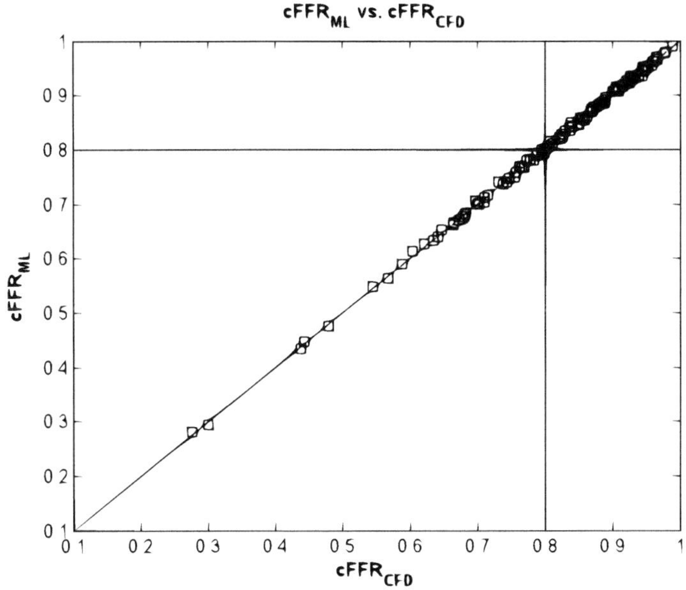

Figure 4.7. Scatterplot of $cFFR_{ML}$ and $cFFR_{CFD}$ (correlation = 0.9994).

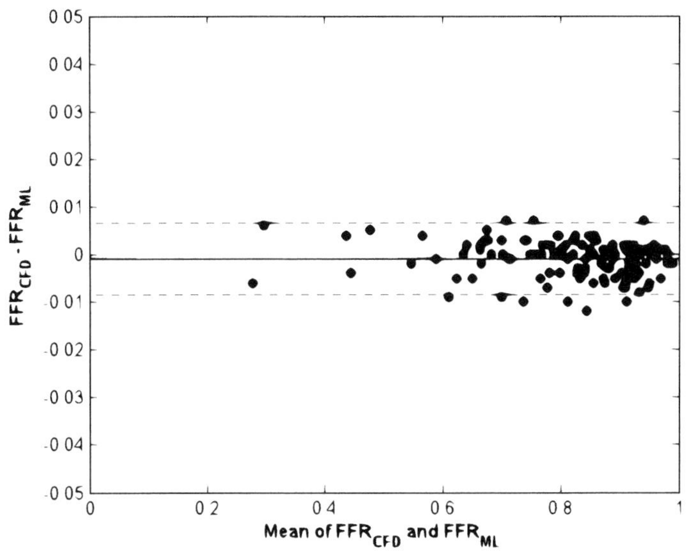

Figure 4.8. Bland–Altman analysis plot comparing $cFFR_{ML}$ and $cFFR_{CFD}$: no systematic bias was found (95% limits of agreement, −0.0085 to 0.0067).

lesions). Thus, $cFFR_{ML}$ had an accuracy of 83.2%, a sensitivity of 81.6% and a specificity of 83.9%. The diagnostic statistics of

Table 4.2 Bin-based comparison between $cFFR_{ML}$ and $cFFR_{CFD}$.

Lesions	$cFFR_{CFD}$ bin	Nr. lesions	Mean difference ± St. dev.
	0.0–0.6	8	0.001 ± 0.004
	0.6–0.7	14	−0.001 ± 0.004
All	0.7–0.8	23	0.000 ± 0.0040
	0.8–0.9	40	−0.001 ± 0.004
	0.9–1.0	40	−0.002 ± 0.003
	0.0–0.6	6	0.000 ± 0.005
	0.6–0.7	10	−0.003 ± 0.004
LAD	0.7–0.8	16	0.000 ± 0.004
	0.8–0.9	27	−0.002 ± 0.004
	0.9–1.0	20	−0.002 ± 0.003
	0.0–0.6	2	0.002 ± 0.003
	0.6–0.7	1	0.002 ±0.000
LCx	0.7–0.8	4	0.001 ±0.003
	0.8–0.9	5	−0.000 ±0.003
	0.9–1.0	11	−0.003 ±0.003
	0.0–0.6	0	–
	0.6–0.7	3	0.003 ± 0.001
RCA	0.7–0.8	3	0.000 ± 0.004
	0.8–0.9	8	0.001 ± 0.002
	0.9–1.0	9	0.001 ± 0.002

both $cFFR_{ML}$ and $cFFR_{CFD}$ are displayed in Table 4.3, alongside their 95% confidence intervals. Overall, the correlation between $cFFR_{ML}$ (mean value of 0.814 ± 0.135) and invasive FFR (mean value of 0.814 ± 0.135) was 0.729 ($p < 0.001$). In Fig. 4.9 and Fig. 4.10 the scatter plots of $cFFR_{ML}$ and $cFFR_{CFD}$ versus invasive FFR, and respectively the corresponding Bland–Altman analyses are displayed. The ROC (receiver-operator characteristic) curves of $cFFR_{ML}$ and $cFFR_{CFD}$ are displayed in Fig. 4.11, both with an AUC (area under the curve) of 0.90.

Finally, in Fig. 4.12 a representative example of a $cFFR_{ML}$ color coded coronary tree is displayed, with invasive FFR = 0.71 and $cFFR_{ML} = 0.68$.

4.1.4 Discussion

In this section a machine learning based approach for near real-time FFR prediction has been introduced, that can be in-

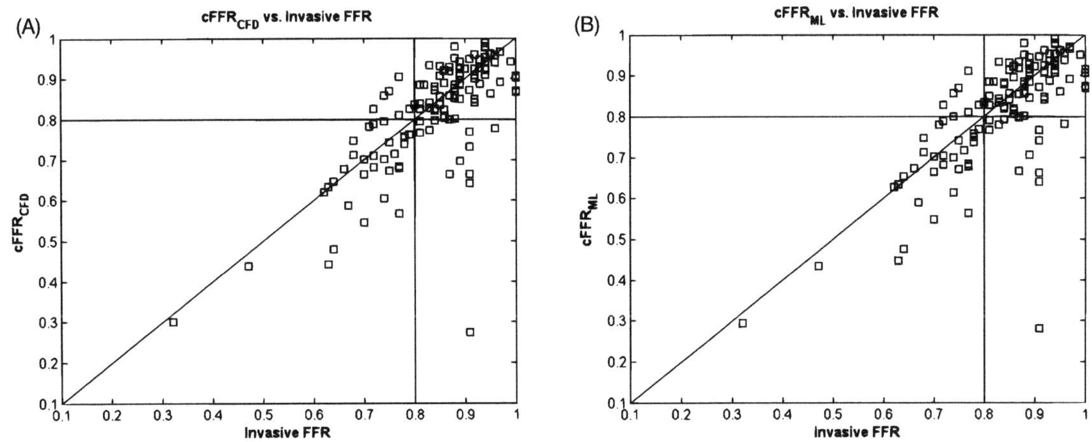

Figure 4.9. (A) Scatterplot of $cFFR_{CFD}$ and invasive FFR (correlation = 0.725); (B) Scatterplot of $cFFR_{ML}$ and invasive FFR (correlation = 0.729).

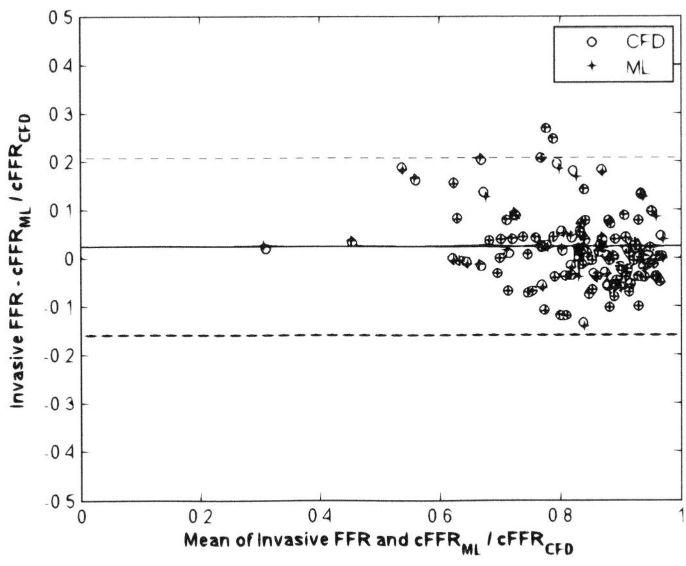

Figure 4.10. Bland–Altman analysis plot comparing $cFFR_{CFD}$ and $cFFR_{ML}$ vs. invasive FFR ($cFFR_{CFD}$ 95% limits of agreement, −0.159 to 0.207; $cFFR_{ML}$ 95% limits of agreement, −0.159 to 0.206).

tegrated in a workstation based solution in the clinical routine. Given the computational and hardware requirements, the approach could be integrated directly into a routine clinical setting. It is relying solely on routine CCTA imaging data and the classification accuracy of 83% is in the range of results reported in the past

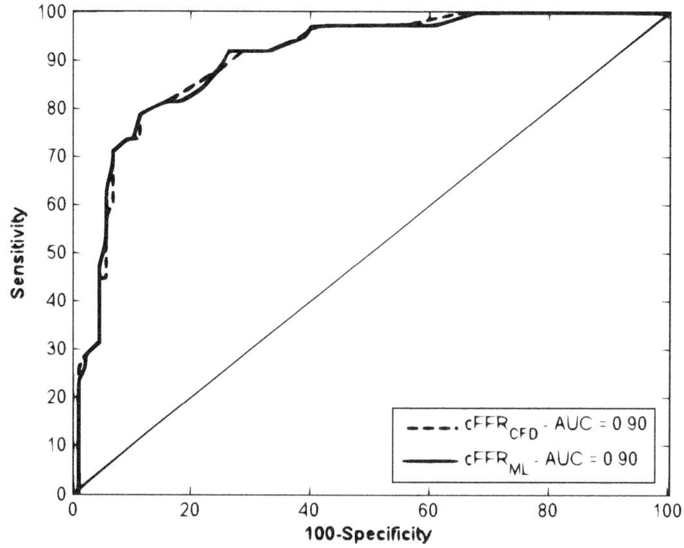

Figure 4.11. Receiver operating characteristic (ROC) curves of $cFFR_{CFD}$ and $cFFR_{ML}$ for 125 lesions.

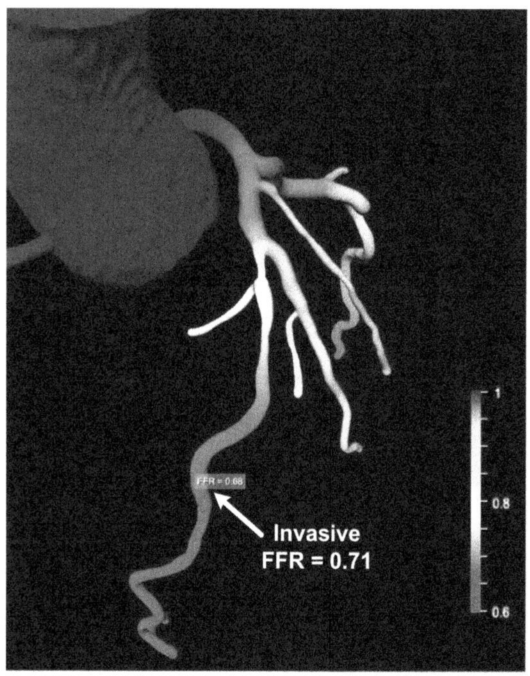

Figure 4.12. Case example of a patient-specific coronary tree with $cFFR_{ML}$ color coded coronary tree: invasive FFR $= 0.71$ and $cFFR_{ML} = 0.68$.

Table 4.3 Diagnostic statistics for patient-specific dataset (N=125); a positive event, i.e. functionally significant lesion, is defined by invasive FFR $\leq$ 0.80. $cFFR_{CFD}$ diagnostic statistics are based on the CFD algorithm, $cFFR_{ML}$ diagnostic statistics are based on the ML algorithm.

Measure	$cFFR_{CFD}$	$cFFR_{ML}$
True positive	31	31
False positive	14	14
True negative	73	73
False negative	7	7
Sensitivity	81.6% (66.6%–90.8%)	81.6% (66.6%–90.8%)
Specificity	83.9% (74.8%–%90.1)	83.9% (74.8%–%90.1)
PPV	68.9% (54.3%–%80.2)	68.9% (54.3%–%80.2)
NPV	91.2% (83.2%–%95.7)	91.2% (83.2%–%95.7)
Accuracy	83.2% (75.6%–%88.7)	83.2% (75.6%–%88.7)
Correlation	0.725	0.729
Mean $\pm$ St. dev.	0.814 $\pm$ 0.135	0.815 $\pm$ 0.135

for non-invasive FFR computation methods: 73% to 85% [322–324, 326,328,346,347]. All of these past studies have demonstrated an incremental value of non-invasively determined FFR over quantitative or visual stenosis grading on CCTA data. Thus, such a solution can further enhance the gatekeeper role played by CCTA for excluding patients from invasive catheterization. Past approaches assessing the hemodynamic significance of CAD from anatomical features alone had limited success [351,352]. The herein proposed approach though relies on a comprehensive set of anatomical characteristics describing not only the lesion, but the entire coronary tree. Further, relying on a state-of-the-art deep learning based approach, and using solely anatomical features for predicting patient-specific hemodynamics can lead to diagnostic statistics that are comparable to those obtained by solving complex hemodynamics equations.

4.1.4.1 Use of synthetic data

Specifically, the key ingredients of the approach are the generation of a very large training database, and the definition of features that are more relevant for coronary hemodynamics. Ideally, the training should be performed based on patient-specific models reconstructed from CCTA images, thus taking into consideration the natural variation in the coronary anatomy across a large number of individuals, and the resulting range of invasive FFR values for each stenosis. However, both the costs and the effort to set up such a large database are prohibitive. To mitigate this aspect, only synthetically generated anatomical models of the coronary tree were employed during the training phase, and generated the corresponding ground truth FFR values using a previously validated blood flow model. The generation of the synthetic database follows a parameterization approach relying on anatomical features of the arterial coronary tree, which enables the generation of various configurations like serial stenoses, three-vessel disease, bifurcation stenoses, diffuse stenoses or rare pathologies. As a result of the training process, the deep learning model encodes the mapping between the geometric characteristics chosen as features and the measure of interest, here FFR, computed by the blood flow model. The anatomical characteristics of the patient-specific data used to define the test set were well within the ranges of values observed in the synthetic training dataset. Given the generic nature of the approach, other coronary hemodynamic measures of interest, like rest PdPa [353], CFR (coronary flow reserve), IFR (instant wave-free ratio) [354], BSR / HSR (basal / hyperemic stenosis resistance) [355,356], wall shear stress [357], etc, may be employed as ground truth during the training process. Moreover, the purely lumen based features defined herein may be augmented with other features related to coronary hemodynamics, e.g. plaque characteristics and patient clinical history, which are important not only for the functional assessment of a lesion but also for its vulnerability in time [358]. Depending on the quantity of interest, a different blood flow model may be employed for defining the ground truth values. Herein, a reduced-order blood model was used, but full-order models may be employed since the time required to build the training database does in general not represent an issue. We note also that the diagnostic performance of $cFFR_{ML}$ is closely linked to the diagnostic performance of the blood flow model employed during the training process. If features are properly selected and the datasets are large enough, the diagnostic statistics of the machine learning model will be indiscernible from those of the blood flow model. Since cFFR is computed at all centerline locations in the anatomical model, virtual pullback curves can be easily extracted, displaying the cFFR variation from a distal location up to the coronary ostium. Pullback curves are specifically useful to distinguish between focal and diffuse lesions and to evaluate the hemodynamic significance of serial lesions independently.

4.1.4.2 Limitations

Several limitations need to be mentioned in the context of the study. As shown in the previous section, the deep learning model was able to almost perfectly reproduce the output of the hemodynamic model, and, thus the limitations are mainly defined based on the latter. First, the patient-specific test dataset was not large enough to allow for a generalization of the diagnostic statistics, and subsequent clinical studies are required for further validation. For example, herein, almost all FFR measurements had a value between 0.7 and 1.0 (FFR < 0.7 was found in only 11 lesions). Nevertheless, the most challenging lesions in terms of diagnosis are those with an FFR value close to the cut-off point of 0.8. A recent study [347] indicated that for the threshold of 0.87 and 0.74 the measures of NPV and PPV respectively were both larger than 90%. Thus a hybrid revascularization strategy may be defined, whereas a cFFR value greater than 0.87 may be used for PCI deferral, a cFFR value smaller than 0.74 may be employed for treatment indication, whilst the remaining lesions may be assessed invasively. The overall agreement with invasive FFR would in this case be higher than 95% in terms of accuracy, and could potentially eliminate the requirement for invasive assessment in around half of the patients. Secondly, the parameters in the CFD model are based on physiological observations, which should be validated extensively. For example, allometric scaling laws are employed to determine the flow rate in each branch, the hyperemic response (in case of microvascular dysfunction, pharmacological vasodilators gave a limited effect, which in turn may lead to an overestimation of the lesion severity), population average blood density and viscosity, etc. Since the latter were set to constant values in the synthetic and patient-specific hemodynamic computations, they have not influenced the correlation between the blood flow and the machine learning model. Nevertheless, these properties can be introduced as additional input variables/features and accounted for using a proper sampling in the training database. Furthermore, collateral vessels were not taken into account. These can alter the blood flow characteristics, especially if severe lesions are present. The synthetic anatomical models represent many common pathological coronary configurations. Nevertheless, the 12 000 synthetic geometries do not address all types of rare pathologies, like coronary aneurysms or anomalous coronary artery origin. Moreover, smooth stenotic shapes were employed which do not reflect the noisy shape variations in stenoses reconstructed from medical images. A future activity will focus on the generation of more realistic stenotic shapes. While by employing the herein proposed approach the cFFR values can be computed in near real-time from an existing anatomical model, a remaining limitation for its routine clinical usage is the time required to reconstruct the anatomical model, which may vary between 10 and 60 minutes. A pre-clinical study employing the same tool and algorithms for the anatomical model generation reported a processing time of 37.5 ± 13.8 minutes [328]. Thus, to mitigate this limitation, an important future activity will focus on the further automation of this step, and the minimization of user interaction.

4.2 Meta-modeling of atrial electrophysiology

Many heart diseases manifest as abnormal rhythms due to disorders that affect the cardiac electrical system. The underlying cause of the disease can be different and requires specific interventions. A range of therapy choices are available, including medical therapy with antiarrhythmic drugs, implantation of pacemakers/defibrillators, ablation of the foci of irregular electrical activity, or surgery. However, many challenges remain open today in patient selection (what patient should undergo what therapy), therapy planning and guidance.

A large body of literature documents steady progress in the understanding of cardiac electrophysiology (EP), thanks to experimental studies in cells, tissues, in animals, and in humans. In particular, significant advances have been made in the understanding of mechanisms underlying heart diseases such as atrial fibrillation (AFib). Atrial fibrillation is a defect of the electrical system of the heart that manifests in irregular contraction of the atria. This causes reduced efficiency of the pumping action of the heart and elevated risks of blood clot formation in the atria – potentially leading to stroke. Various treatment options are available, including pharmacological, surgical and non-surgical approaches (such as ablation). Selection of the optimal course of treatment as well as planning of the optimal delivery of the treatment (for instance in the case of catheter ablation) are still open clinical challenges. Computational modeling of cardiac EP and arrhythmia has emerged as a complimentary approach to experimental studies for the understanding of the mechanisms producing abnormal cellular activation patterns, with the potential of enabling non-invasive, cost effective and individualized analysis of the electrophysiological state and evolution of the heart. Recent modeling studies achieved remarkable fidelity in reproducing observable electrophysiological phenomena, and hold the promise of a new generation of tools for individualized patient care.

Specialized cellular models have been developed for the description of electrophysiology in different tissues in the heart, including the behavior of ventricular and atrial myocardium, as well as tissue belonging to the conduction system of the heart (sinoatrial node, atrio-ventricular node, His bundle and Purkinje system). Basic models of cellular electrophysiology stem from the seminal work of Hodgkin and Huxley [12], and describe ionic interactions across the cell membrane. More recent models include a detailed representation of the biological phenomena controlling ion channels, ion interactions and underlying cell function (see section 1.2.1). Such models are computationally demanding due to the numerous and coupled algebraic and ordinary differential equations governing the dynamics of different ionic channels and gating variables. As an example, the Courtemanche–Ramirez–Nattel (CRN) atrial cell model [359] has 35 static parameters, 12 ionic channels, and 21 ordinary differential equations for gating variables and ionic concentrations. A diagram showing the different components of the CRN model is shown in Fig. 4.13. While the descriptive power of the model can potentially greatly increase, as more aspects of the cell electrophysiology are explicitly captured, additional challenges arise from the uncertainty associated to the model parameters. Since many of the microscopic phenomena described by the model cannot be directly observed by non-invasive techniques, the parameters may have to be chosen as representative of the average behavior of populations or groups of persons, based for instance on extrapolation from data acquired from animal studies or ex-vivo analysis. Even when the phenomena described by the model can be observed to a certain extent (either directly or indirectly), the joint estimation of multiple model parameters for the definition of a patient-specific model presents significant methodological challenges (see chapter 5).

This motivates the design of models that are computationally efficient (potentially enabling real-time simulations), reliable in capturing the main aspects of cardiac electrophysiology, and whose parameters can easily be estimated, directly or statistically, from clinical data. Several studies proposed models with these properties [64], generally providing the representation of the macroscopic behavior of the cell or tissue rather than a detailed description of the underlying ionic phenomena. The focus on macro-scale dynamics allows the reduction of the number of model parameters, which are directly related to the features of measurable signals (e.g. the shape of the action potential or ECG measurements). A potential disadvantage, however, is that the recovered dynamics may lack important features linked to the underlying physics, such as distinct time scales for the depolariza-

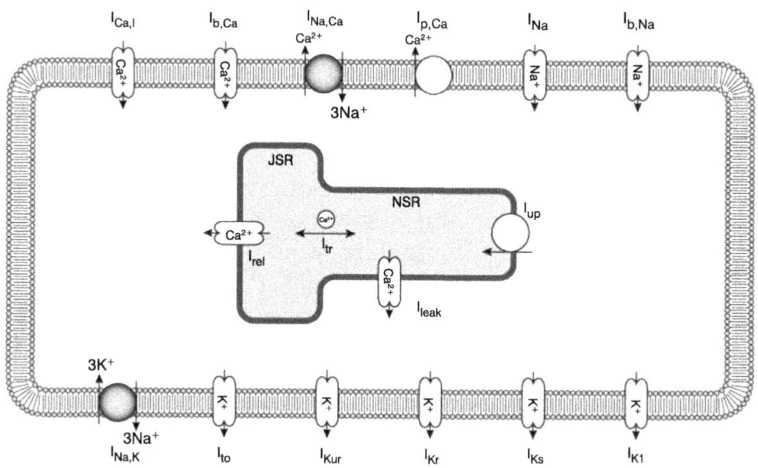

Figure 4.13. Diagram of the ionic channels in the CRN atrial cell model *(source: CellML, https://models.cellml.org/exposure/0e03bbe01606be5811691f9d5de10b65).*

tion and repolarization phase, and the dependence of action potential duration and conduction velocity on heart rate. Moreover, phenomenological models have not been developed for certain cells and tissues, such as human atrial cells.

In this scenario, great benefit would come from new and improved techniques for the design of reduced models of human cardiac electrophysiology, able to reproduce important tissue-level characteristics while maintaining a reduced number of parameters to allow for efficient estimation from clinical data.

Model reduction techniques relying on statistical learning (often referred to as meta-modeling) have been used in particular in the chemometrics community. The idea is to derive a statistical model that is able to predict the output of non-linear computational models which have complex dynamics, while being expressed with fewer parameters and being computationally extremely efficient [333]. In the context of computational electrophysiology, Sobie [360] first provided an example of this approach, using a partial least squares regression (PLSR) model to predict the outputs of mechanistic models of ventricular action potential. This work demonstrated that multivariate linear regression on the input parameters allows for a recovery of quantities of interest, such as action potential duration, with good correlation when compared to the results of the mechanistic models. Mansi et al. [361] also used a data-driven approach with manifold learning techniques to predict the sarcomere force computed from a cardiac myofilament model.

Motivated by recent results in meta-modeling, we explore here a reduced model of atrial cellular electrophysiology, based on state-of-the-art mechanistic models and data reduction techniques. The reduced model allows for an efficient computation, capturing the dynamics of the original biophysically detailed model with less input parameters. This chapter also illustrates the use of such reduced model based on statistical learning for multiscale modeling of cardiac electrophysiology. In the following we first focus on the CRN atrial cell model (section 4.2.1.1). We consider the dynamics of the action potential described by that model and build a lower-dimensional manifold using manifold learning techniques such as principal component analysis (PCA), or locally linear embedding (LLE) (section 4.2.1.2). Next, we learn a forward regression model of the action potential dynamics, given a set of CRN model parameters (section 4.2.1.3). Finally we integrate this reduced cellular model into tissue-level EP modeling (section 4.2.1.4).

We show in section 4.2.2 that the dimension of the action potential manifold can be reduced significantly while still reproducing the output of a non-linear system. On the other hand, this drastically improves performance of the atrial tissue-level electrophysiology modeling, by reducing computational cost by up to two orders of magnitude. Section 4.2.3 discusses these results and proposes future considerations. The content of this section is based on work published in [362].

4.2.1 Methods

4.2.1.1 Atrial electrophysiology models

The Courtemanche–Ramirez–Nattel (CRN) cell model [359] was developed based on human atrial cell data and was validated for both tissue- [363] and organ- [364] level electrophysiology simulations. The CRN model considers 12 ionic channels with 21 variables including voltages, gating variables for the transmembrane currents and calcium (Ca^{2+}) ion release current, Ca^{2+} concentrations and intracellular sodium Na^+ and potassium (K^+) ion concentrations. The governing equation is

$$\frac{dv}{dt} = -I_{ion} - I_{stim} \tag{4.2}$$

with

$$\begin{aligned} I_{ion} = I_{Na} + I_{K1} + I_{to} + I_{Kur} + I_{Kr} + I_{Ks} + I_{Ca,L} + I_{p,Ca} \\ + I_{NaK} + I_{NaCa} + I_{b,Na} + I_{b,Ca} \end{aligned} \tag{4.3}$$

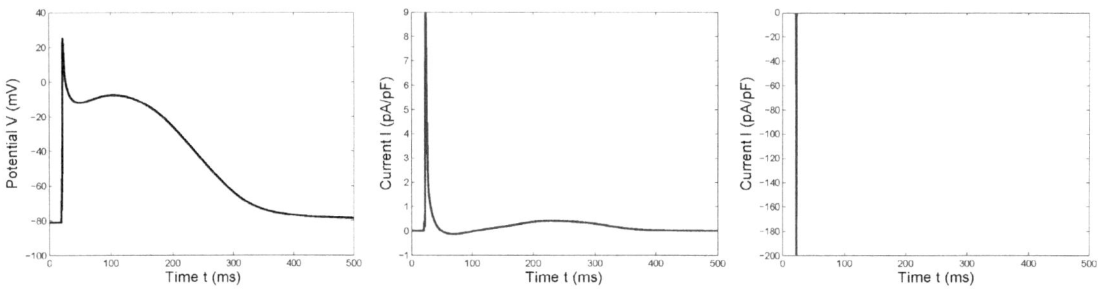

Figure 4.14. Left: Action potential of the CRN model with parameters as in [359]; Middle: Current $I_{ion}-I_{Na}$; Right: Current I_{Na}.

where the terms in (4.3) represent different components of the global ionic current generated by the model. An example of action potential profile and ionic currents produced by the model is shown in Fig. 4.14.

The large number of parameters of the CRN model determines a high-dimensionality manifold Ω_{AP} of parameter tuples generating $v(t)$. In general, this makes it challenging to estimate patient-specific models, because individual parameters may not be directly observable, and different parameters may correlate with each other.

4.2.1.2 Learning the action potential manifold for dimensionality reduction

To reduce the dimensionality of the manifold Ω_{AP}, a powerful tool is manifold learning. The underlying assumption is that a reduced subset of model parameters is sufficient to capture the most significant modes of variation of the action potential $v(t)$, as observed in a database of realizations.

Let us assume that n observations of $v(t)$ are available. The i-th observation $v^i(t)$ is generated by solving (4.2) with a unique set of parameters $\boldsymbol{\theta}^i$, and m snapshots are gathered in the observation vector $\mathbf{v}^i = [v^i(t_1), \cdots, v^i(t_m)] \in \mathbb{R}^m$. The AP manifold is then defined as the subspace of $\mathbb{R}^m$, containing the n observations, and it is represented by the $n \times m$ observation matrix $\mathbf{Y} = [(\mathbf{v}^1)^T, \cdots, (\mathbf{v}^n)^T]^T$. To uncover intrinsic structures of the AP manifold, linear or nonlinear dimensionality reduction can be applied, looking for an embedded manifold within Ω_{AP}, with a significantly lower dimension. As examples of both approaches, we describe the application of principal component analysis (PCA) [365], and of locally linear embedding (LLE) [366].

PCA projects the high-dimensional space Ω_{AP} onto the reduced space Ω_{AP}^{pca} spanned by principal components $\mathbf{z}_l$ ($l \in \{1, \cdots, q\}$) that are orthogonal (uncorrelated) and maximize the observed covariance. The principal components are computed as $\mathbf{Z} = \mathbf{Y}_{zs}\mathbf{W}$ where $\mathbf{Y}_{zs}$ is the observation matrix converted in z-scores, $\mathbf{Z} = [\mathbf{z}_1, \cdots, \mathbf{z}_q] \in \mathbb{R}^{n \times q}$, $\mathbf{W} = [\mathbf{w}_1, \cdots, \mathbf{w}_q] \in \mathbb{R}^{m \times q}$. The $\mathbf{w}_l$'s are the eigenvectors of the covariance matrix $\mathbf{Y}_{zs}\mathbf{Y}_{zs}^T$ in decreasing order of corresponding eigenvalues. Given an observation by the PCA coefficient row vector $\mathbf{z}_{pca}$, the reconstruction from space Ω_{AP}^{pca} to space Ω_{AP} is by $\mathbf{v} = \bar{\mathbf{y}} + \mathbf{z}_{pca}\mathbf{W}^T \odot \boldsymbol{\sigma}_Y$. Here $\bar{\mathbf{y}}$ and $\boldsymbol{\sigma}_Y$ are the mean and standard deviation of $\mathbf{Y}$, and $\odot$ represents an element-wise product.

LLE seeks a projection of data onto a low-dimensional space Ω_{AP}^{lle} by a neighborhood-preserving mapping. The standard LLE algorithm is performed in three steps. First, k nearest neighbors are identified for each data point $\mathbf{v}^i$, based on a given distance metric; then, the barycentric coordinates w_{ij} of $\mathbf{v}^i$ relative to its k neighbors are computed; finally, the embedding coordinates $\mathbf{z}_{lle}^i \in \Omega_{AP}^{lle}$ are found by minimizing the global embedding cost function $\Phi(\mathbf{z}_{lle}) = \sum_i |\mathbf{z}_{lle}^i - \sum_j w_{ij}\mathbf{z}_{lle}^j|^2$, with given weights w_{ij}. To reconstruct an observation, mapping $\mathbf{z}_{lle}$ from space Ω_{AP}^{lle} to space Ω_{AP}, first the k nearest neighbors in space Ω_{AP}^{lle} are identified, and then the observation is reconstructed as the linear combination of its nearest neighbor, using the same weights.

4.2.1.3 Statistical learning

We define a data-driven model of action potential dynamics, generating an AP profile $\mathbf{v} \in \mathbb{R}^m$ given a sample $\boldsymbol{\theta} \in \mathbb{R}^p$ of the CRN model parameters. The model writes $\mathbf{v}_{dr} = \mathbf{f}(\boldsymbol{\theta})$, $\mathbf{f} : \mathbb{R}^p \rightarrow \mathbb{R}^q$, and it is designed to predict the embedding coordinates described in section 4.2.1.2, which are then used to reconstruct the AP profile $\mathbf{v}$. Depending on the chosen embedding, $\mathbf{v}_{dr} = \mathbf{z}_{pca}$ or $\mathbf{v}_{dr} = \mathbf{z}_{lle}$. In the following, the parameter values $\boldsymbol{\theta}$ are standardized using z-scores when used as input for the regression model. To simplify notation, we will not use a new symbol for the standardized parameter values.

The regression problem is split into two parts. In the first step, the relationship between CRN model parameters and handcrafted features of the action potential profile are estimated. These phenomenological features are then used as additional input for the second regression step, estimating the embedding coordinates of the AP, to increase the accuracy of the overall prediction.

The additional features used in the regression are peak voltage (V_{peak}), resting membrane potential (V_{rest}), and action potential duration (APD). The first model then reads $\mathbf{f}_1 : \boldsymbol{\theta} \mapsto [V_{peak}, V_{rest}, APD]$, and can be estimated for instance using the projection pursuit regression (PPR) method [367]. The central idea of PPR is to define auxiliary features by computing linear combinations of the model inputs in an optimal direction; and then model the output as a nonlinear function of the auxiliary features. Let f_1 represent any of the outputs of $\mathbf{f}_1$; using PPR, the model has the form:

$$f_1(\boldsymbol{\theta}) = \sum_{j=1}^{s} g_j(\boldsymbol{\theta}\boldsymbol{\omega}_j), \tag{4.4}$$

where $\boldsymbol{\omega}_j$ is a unit column vector in $\mathbb{R}^p$, and function g_j is estimated as part of the model creation. The model can in fact be built in an iterative manner: given a projection direction $\boldsymbol{\omega}_k$, apply a univariate smoother to obtain an estimation of g_k by minimizing the discrepancy between model and data; given a function g_k, use Gauss–Newton search to find the optimal updated weights $\boldsymbol{\omega}_k$ minimizing the fitting error. The number of ridge terms s is determined by a backward deletion procedure.

The second step of the statistical learning procedure is to include predicted phenomenological features of the action potential as model inputs together with the CRN model parameters, for estimating the embedding coordinates $\mathbf{v}_{dr}$. This step estimates a model $\mathbf{f}_2 : [\boldsymbol{\theta}, \mathbf{f}_1(\boldsymbol{\theta})] \mapsto \mathbf{v}_{dr}$, which can be built for instance using multivariate adaptive regression spline (MARS [365]), or the PPR method as in the first step. MARS is a non-parametric regression method extending linear regression by fitting linear or cubic splines to capture data non-linearity. Let f_2 represent any of the outputs of $\mathbf{f}_2$; using MARS, the model has the form:

$$f_2(\boldsymbol{\theta}, \mathbf{f}_1(\boldsymbol{\theta})) = \beta_0 + \sum_{j=1}^{J} \beta_j h_j(\boldsymbol{\theta}, \mathbf{f}_1(\boldsymbol{\theta})), \tag{4.5}$$

where $h_j(x)$ is a function of the form $\max(x-c, 0)$ or $\max(c-x, 0)$ or their products, with c being a constant. The model is estimated by a greedy approach fitting the splines to the training data, and then a backward deletion pass is performed on the splines to minimize over-fitting.

This multi-step regression approach can be further extended by iteratively using the estimated components of $\mathbf{v}_{dr}$ as additional inputs. More precisely, $\mathbf{v}_{dr}(1)$ can be estimated using $[\boldsymbol{\theta}, \mathbf{f}_1(\boldsymbol{\theta})]$ as inputs; $\mathbf{v}_{dr}(2)$ can be estimated using $[\boldsymbol{\theta}, \mathbf{f}_1(\boldsymbol{\theta}), \mathbf{v}_{dr}(1)]$, and similarly for the following components of $\mathbf{v}_{dr}$.

4.2.1.4 Application to tissue-level cardiac EP modeling

The monodomain model described in section 1.2.2 describes the temporal and spatial distribution of the transmembrane potential $v(t)$ in cardiac tissue. In particular, when coupled with the CRN cell model, the mono-domain equations describe the electrical activation of human atrial tissue. The disadvantages of using full-order cellular models are the computational complexity associated to the numerical solution of the underlying ordinary differential equations, and the challenges of estimating an often large number of patient-specific model parameters. A reduced order, data driven model of the action potential time profile allows for both a significant gain in computational efficiency, by removing the need to solve the model equations, and simplifies the parameter estimation problem, by enabling fast model evaluation, and therefore a more efficient optimization procedure.

A large database of sample transmembrane potential profiles can be generated on the reference anatomical model of the human atria, with varying values of the CRN model parameters $\boldsymbol{\theta}$, by numerically solving the monodomain equation. For each sampled set of CRN model parameters, the computed ionic current $J_{ion}(t)$ is recorded at a pre-defined measurement site in the computational domain. For the sake of simplicity, we hereby consider periodic electrical activity and focus on a single heart beat. The database used for regression is then built by collecting observations of the action potential defined as $v(t^i) = v(t^0) + \sum_{j=1}^{i} J_{ion}(t^i)dt$. The regression approach described in the previous section allows for the definition of an action potential profile model as a function of the cellular model parameters. In the following, we refer to the estimated action potential profile as $\mathbf{v}_{ref}$; and to the corresponding ionic current as $\mathbf{J}_{ref}$.

The estimated action potential profile can then be used in combination with a numerical solver of the monodomain equation, to compute the electrical activation of the target cardiac tissue for a new set of model parameters. To achieve this, the ionic current term in Eq. (1.5) is replaced by the estimated $\mathbf{J}_{ref}$, after the action potential is triggered, and for each time frame in which the monodomain equation is solved.

To identify the time at which the action potential is generated at each location in the computational model, Eikonal models can be used as a very efficient option. However, as discussed in section 1.2.2, Eikonal model do not allow modeling pathological phenomena like arrhythmias. Alternatively, phenomenological models can be used to simulate the temporal and spatial evolution of the transmembrane potential in the cardiac tissue. One option is to use the monodomain model based on the Mitchell–Schaeffer

Table 4.4 Parameters in the CRN model.

Parameter	Definition	Baseline value
g_{Na}	Maximal fast inward Na^+ current (I_{Na}) conductance	7.8 nS/pF
g_{K1}	Maximal inward rectifier K^+ current (I_{K1}) conductance	0.09 nS/pF
g_{to}	Maximal transient outward K^+ current (I_{to}) conductance	0.1652 nS/pF
g_{Kr}	Maximal rapid delayed rectifier K^+ current (I_{Kr}) conductance	0.0294 nS/pF
g_{Ks}	Maximal slow delayed rectifier K^+ current (I_{Ks}) conductance	0.129 nS/pF
$g_{Ca,L}$	Maximal L-type inward Ca^{2+} current ($I_{Ca,L}$) conductance	0.1238 nS/pF
$g_{b,Ca}$	Maximal background Ca^{2+} current ($I_{b,Ca}$) conductance	0.00113 nS/pF
$g_{b,Na}$	Maximal background Na^+ current ($I_{b,Na}$) conductance	0.000674 nS/pF
$I_{NaK(max)}$	Maximal $Na^+ - K^+$ pump current (I_{NaK})	0.60 pA/pF
$I_{NaCa(max)}$	Maximal Na^+/Ca^{2+} exchanger current (I_{NaCa})	1600 pA/pF
$I_{p,Ca(max)}$	Maximal sarcoplasmic Ca^{2+} pump current ($I_{p,Ca}$)	0.275 pA/pF
$g_{Kur(max)}$	Scale factor of ultrarapid delayed rectifier K^+ current (I_{Kur})	1

cellular model [71], described in section 1.2.2. For each location in the computational domain, the time frame $t_{upstroke}$ in which action potential generation is triggered is computed using this approach. During each following time frame $t^i = t_{upstroke} + i \cdot dt$, with dt representing the time step used for the numerical solution of the monodomain equation, the regression cellular model is used as follows:

$$dt\ J_{ion}(\mathbf{x}, t^i) = \mathbf{v}_{ref}(i) - v(\mathbf{x}, t^{i-1}) \approx \mathbf{v}_{ref}(i) - \mathbf{v}_{ref}(i-1) = dt\ \mathbf{J}_{ref}(i). \tag{4.6}$$

Analogously, at the end of the action potential the fully coupled monodomain problem with MS cellular model is solved, to monitor the generation of AP in the next heart beat.

4.2.2 Experiments and results

4.2.2.1 Model parameter selection and sampling

Without loss of generality, we focus in the following on a subset of the parameters of the CRN model. More precisely, we consider 12 out of the 35 parameters used by the model, as reported in Table 4.4. The rationale behind this choice is that the meta-modeling approach can be applied regardless of the number of considered model inputs; moreover, the action potential dynamics produced by the CRN model when varying the 12 selected parameters is very rich, capturing a large variety of AP patterns. As shown in Fig. 4.15, observations generated by randomly sampling the 12 parameters in a normal distribution with mean as noted in Table 4.4, and standard deviation SD = 0.3, have similar variability as observations generated by sampling all model parameters in the same distribution. An additional advantage is that limiting the number of inputs for the statistical learning is likely to improve the accuracy of the data-driven model.

For the generation of the database of observations used in the regression, samples of the set of 12 selected CRN model parameters were generated by scaling the baseline values listed in Table 4.4 (see also [359]). Random scaling factors were chosen from a log-normal distribution with a mean value of 1. The CRN model governing equation was solved, with a stimulus I_{stim} having an amplitude of −20pA/pF, starting at $t = 10$ ms, and lasting for 1 ms. In the first 500 ms, the transmembrane potential profile $v(t)$ was recorded as output, sampled in time with time step $dt_{snap} = 0.5$ ms.

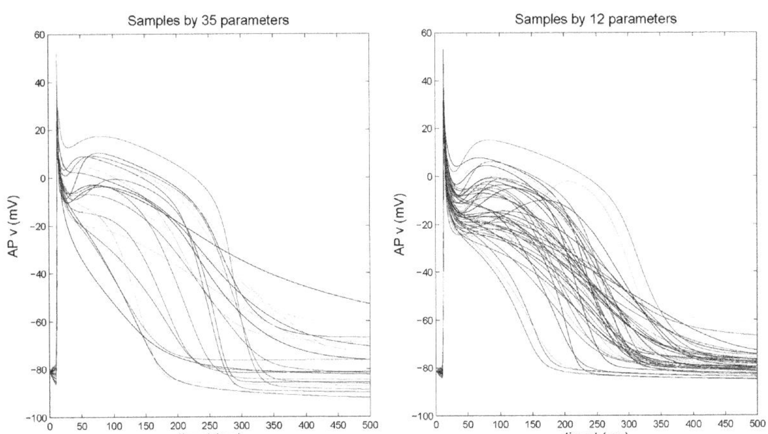

Figure 4.15. Samples with SD=0.3 by different number of parameters.

In the first experiment, the training set consists of 1000 observations extracted from the observation database described above; while the testing set consists of 500 observations from a different database (generated by sampling the CRN model parameters from a different distribution). The goodness of fit of the predicted output is measured in different ways: R^2 value defined as $R^2 = 1 - \frac{SS_{res}}{SS_{tot}}$ with SS_{res} being the residual sum of squares, and SS_{tot} being the total sum of squares, Maximum Amplitude Difference (MAD, in mV) defined by $MAD(y_i, y_j) = |\max(y_i(t)) - \max(y_j(t))|$, and absolute differences between Areas Under the Curves (AUC).

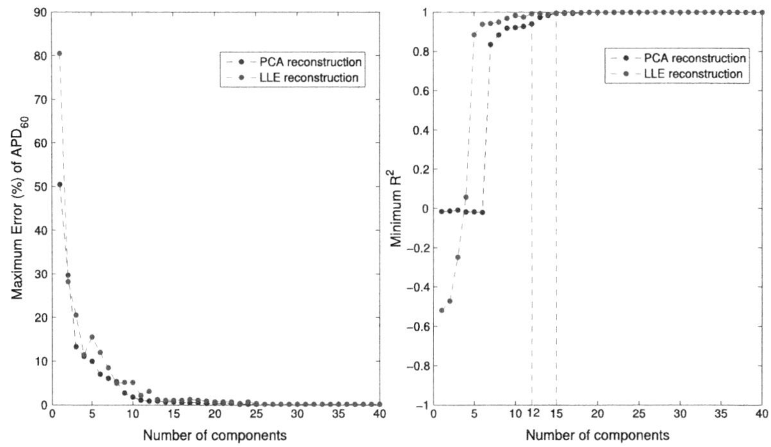

Figure 4.16. Goodness of reconstruction on testing data using PCA and LLE.

In all cases, such metrics are computed with the original data before standardization (using z-scores) or dimensional reduction (using embedding). The resting potential V_{rest} is defined as the potential recorded in the last time frame, i.e. at $t = 500$ ms. The action potential duration APD_{20} (APD_{40}, APD_{60}) is defined as the time from AP onset ($\arg_t \max \frac{dv}{dt}$) to the time at which $v(t) = -20$ mV (-40 mV, -60 mV). For the training dataset in this experiment, the MAD, V_{rest}, APD_{60} variations in mean $\pm$ SD form were 29.2772 ± 8.43309 mV, -77.4807 ± 5.61862 mV, and 269.346 ± 67.45435 ms respectively.

4.2.2.2 PCA versus LLE

The reduced spaces Ω_{AP}^{pca} and Ω_{AP}^{lle} are computed using the training set. To evaluate the performance of the two manifold learning techniques, the accuracy of the reconstructed action potential profile is evaluated by computing the maximum relative error (relative to the mean value in the training database) of APD_{60} and the maximum R^2. These metrics are plotted in Fig. 4.16, on the left and right side respectively. In the plot on the right side, each red (light gray in print version) vertical line denotes the minimum number of components needed for R^2 to be greater than 0.99. For LLE, the number of neighbors k is chosen as $\max(20, n_{comp})$ with n_{comp} being the number of components. The number of neighbors can be optimized through benchmarking, to improve the accuracy of the reconstructed AP profiles; nonetheless, the number of components seems to have the most important effect on accuracy. As shown in the figure, both PCA and LLE require more than 10 components to accurately capture AP dynamics, 15 components

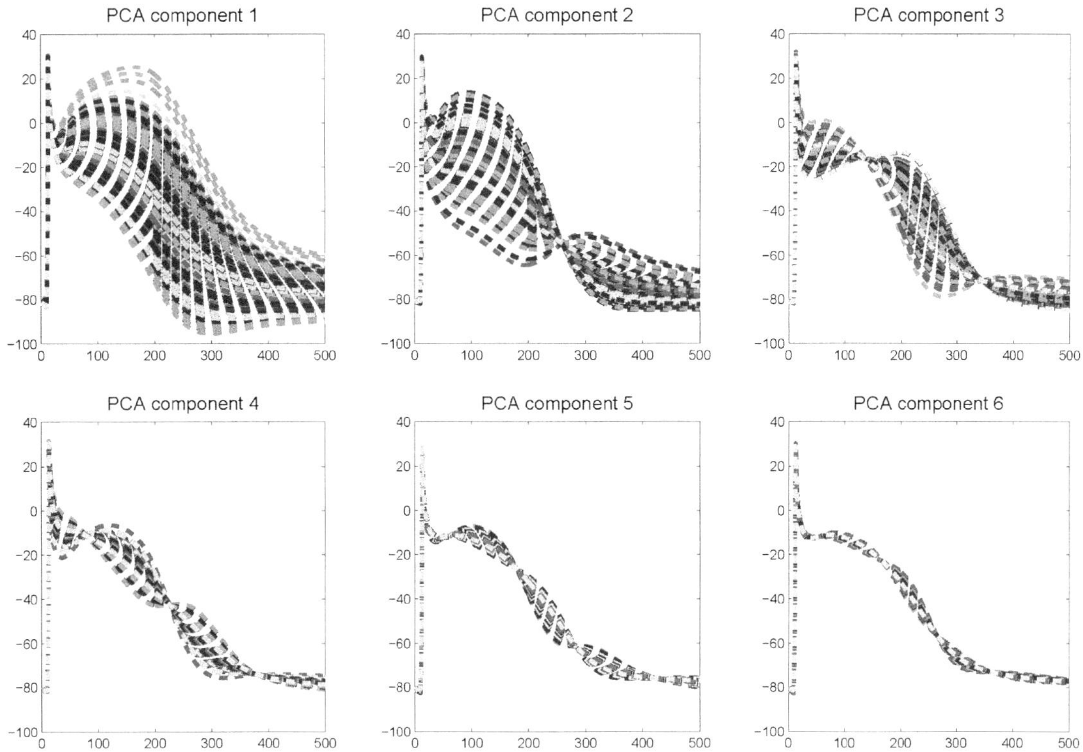

Figure 4.17. Modes of variation of the action potential profile produced by the CRN model, estimated by PCA components.

being sufficient for both. This is a significant reduction in dimensionality (compared with the number of degrees of freedom in the original AP profile, i.e. 1000 as the number of temporal frames). Both manifold learning techniques perform similarly on the considered AP manifold Ω_{AP}, suggesting that non-linearities are not important in this problem. In the following, we only use PCA in virtue of its performance and computational efficiency. Fig. 4.17 illustrates the modes of variations of the AP profile estimated by PCA components. In particular, the first mode captures the AP amplitude, the following modes capture variations of curve concavity during different periods of time.

4.2.2.3 Physical regression model construction

As described in section 4.2.1.3, we use a multi-step regression approach. In the first step, a model is created to predict $[V_{peak}, V_{rest}, APD_{60}, APD_{40}, APD_{20}]$ from the CRN model parameters provided as input. To build the model, three regression methods are tested: partial least squares regression (PLSR), MARS, and PPR. The number of components option is set equal to the number of model inputs p in PLSR. For MARS, the maximum interaction degree is set to 2, and the maximum number of model terms to 80. For PPR, the number of terms in the model is set to 10, and the optimization level is set to re-balance the contributions from each regressor at each step. The computed R^2 values for the outputs estimated by all methods are reported in Fig. 4.18 and Fig. 4.19. Despite its popularity, PLSR produces a model that can only accurately predict V_{peak}, but fails at estimating V_{rest}. Previous work by Sobie [360] reported better PLSR performance for the definition of a model of ventricular electrical activity. However, Fig. 4.18 demonstrates the non-linear complexity of the CRN atrial model. This result is consistent with the sample patterns in Fig. 4.15 and the mode variations represented by PCA. We also find that MARS can accurately predict V_{peak} and V_{rest}, but returns worse results in APD prediction. PPR performs best in this test, with all the R^2 values being above 0.97.

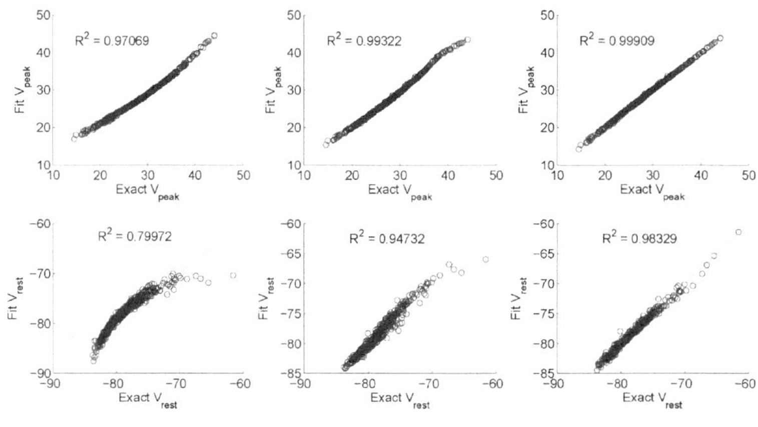

Figure 4.18. Regression on V_{peak} and V_{rest}: PLSR (1st column), MARS (2nd column), PPR (3rd column).

We also study the generalization capability of the PPR model, by evaluating the accuracy of the model predictions for samples of the input model parameters outside the distribution used to generate the training set. In Fig. 4.20, the x-axis represents the standard deviation of the distribution used to sample the input parameters for generating testing data. The y-axis shows the computed R^2 value for the model outputs. In this experiment, R^2 values decrease with increasing SD, but remain higher than 0.9 for SD smaller than 0.4. For SD=0.5, the predictor has notably worse performance. This suggests that the model has a modest general-

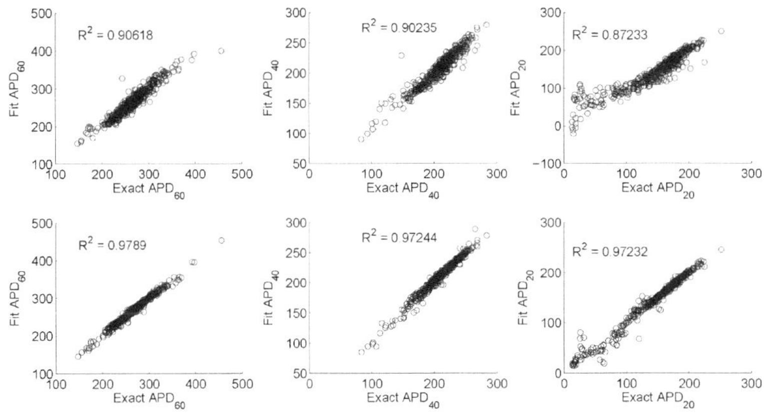

Figure 4.19. Regression on APD: MARS (1st row), PPR (2nd row).

ization capability, and a strategy to control the performance could be based on the definition of multiple specialized models, each one covering a different distribution of the input parameters, for instance tailored to represent pathological cases and their typical input parameter distribution (e.g. the case of atrial fibrillation, in which APD is typically shorter). The corresponding training database of observations could be generated by sampling input parameters in a modified range as in [368]. In the following, we will only consider distributions of input parameters in the range of the published reference values for healthy persons [359].

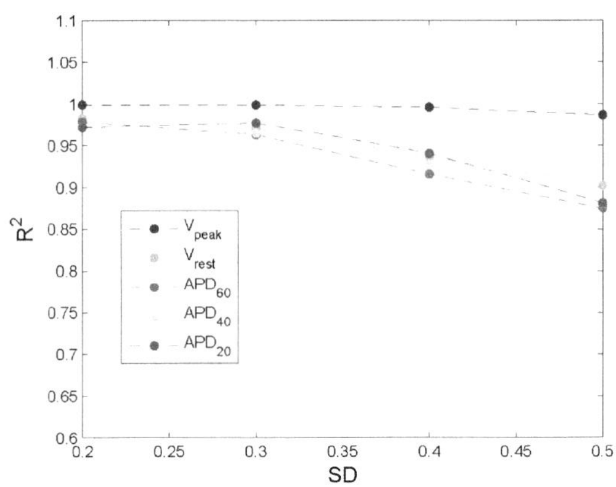

Figure 4.20. Accuracy of the model predictions with increasing standard deviation of the distribution of model parameters used to generate the testing data.

In the second step of model construction, we use outputs from the first model $\mathbf{f}_1 = [V_{peak}, V_{rest}, APD_{60}, APD_{40}, APD_{20}]$ as additional features for a second model $\mathbf{f}_2 : [\boldsymbol{\theta}, \mathbf{f}_1(\boldsymbol{\theta})] \mapsto \mathbf{v}_{dr}$. Further, we also consider a variant of the second model, in which components of $\mathbf{v}_{dr}$ are iteratively estimated, and used as additional features for the regression of the remaining ones. This defines a total of four methods:

$\mathcal{M}_1$ use only $\boldsymbol{\theta}$ as model input;
$\mathcal{M}_2$ use $[\boldsymbol{\theta}, \mathbf{f}_1]$ as model input
$\mathcal{M}_3$ use $[\boldsymbol{\theta}, \mathbf{v}_{dr}(1), \cdots \mathbf{v}_{dr}(i)]$ as model input to predict $\mathbf{v}_{dr}(i+1)$
$\mathcal{M}_4$ use $[\boldsymbol{\theta}, \mathbf{f}_1, \mathbf{v}_{dr}(1), \cdots \mathbf{v}_{dr}(i)]$ as model input to predict $\mathbf{v}_{dr}(i+1)$

We evaluate the four methods using both MARS and PPR for model estimatation. The embedding Ω_{AP}^{pca} is generated as the span of 15 PCA components.

We report in Table 4.5 the fitting errors of different quantities of interest of the action potential profile, when using a model estimated by PPR using the four different methods. All reported errors are computed as the difference between the quantity as extracted from the predicted action potential profile, and the quantity extracted from the ground truth profile; normalized by the mean value of the same quantity in the training database. Small errors in MAD and V_{rest} suggest that the profiles estimated by the model capture correctly the amplitude of the action potential; small errors in APD suggest that the time pattern is properly reconstructed; small errors in AUC indicate a global goodness of fit, with no significant localized discrepancies between the predicted and ground truth action potential profiles. The results indicate that using $\mathbf{f}_1$ as additional input parameter for model estimation reduces the error in APD, but not the error in MAD. Introducing the iterative estimation of the components of $\mathbf{v}_{dr}$ significantly decreases the MAD error. Overall, method $\mathcal{M}_4$ produces the most accurate prediction. An example of the estimated action potential profiles by PPR with $\mathcal{M}_4$ is shown in Fig. 4.21.

The model estimated with MARS achieved comparable performance when using method $\mathcal{M}_4$, and slightly lower performance when using the other methods. MARS has a potential advantage over PPR since it produces faster predictions thanks to a simpler model structure. This can be relevant for applications requiring repeated estimation of the action potential profile, such as parameter estimation for model personalization.

4.2.2.4 Application in tissue-level EP modeling

We tested the application of the regression cellular model to the study of the electrical activation of cardiac tissue. Starting from diagnostic cardiac images, the left atrium and right atrium are automatically segmented using a machine learning approach [31]

Table 4.5 PPR prediction error.

		$\mathcal{M}_1$	$\mathcal{M}_2$	$\mathcal{M}_3$	$\mathcal{M}_4$
$MAD(\%)$	Mean±SD	23.32±16.43	23.25±16.50	0.903±1.214	0.985±1.465
	90-percentile	45.10	45.72	1.749	2.098
$V_{rest}(\%)$	Mean±SD	0.86±0.91	0.74±0.85	0.37±0.50	0.67±0.80
	90-percentile	1.83	1.64	0.81	1.42
$APD_{60}(\%)$	Mean±SD	3.35±3.59	2.66±3.12	2.88±3.41	2.73±3.31
	90-percentile	7.51	6.12	6.59	6.12
$APD_{40}(\%)$	Mean±SD	3.56±4.14	2.84± 3.05	2.89±3.52	2.76±2.841
	90-percentile	8.30	5.98	6.35	5.61
$APD_{20}(\%)$	Mean±SD	5.29±7.02	4.05±4.94	3.72±6.94	3.51±7.47
	90-percentile	12.37	8.62	8.06	7.87
$AUC(\%)$	Mean±SD	1.70±2.06	1.39±1.55	1.74±2.06	1.26±1.58
	90-percentile	3.65	2.99	3.82	2.70

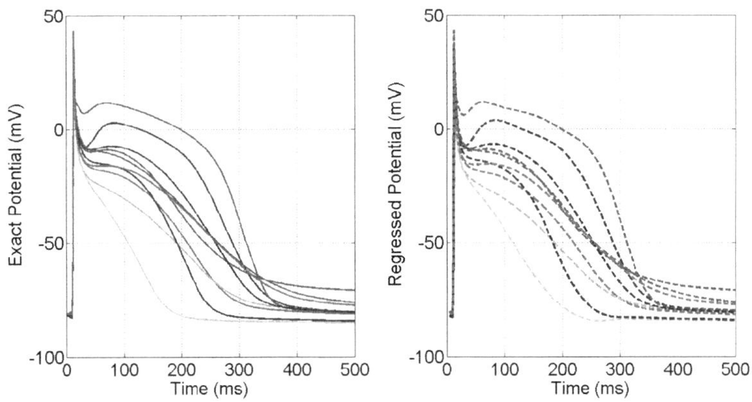

Figure 4.21. AP regression by PPR.

and fused in one surface representing the myocardium. Regional atrial wall thickness is assumed to be uniform. The sino-atrial node region is manually marked in the anatomical model and used to initialize the electrical activation. Computational modeling of tissue-level electrophysiology is based on the monodomain equation, solved with the lattice-Boltzmann algorithm described in section 2.2.1.

As detailed in section 4.2.1.4, the regression cellular model provides a reference action potential $\mathbf{v}_{ref}$ as a time sequence

Table 4.6 Computational cost comparison.

Model	dt	$T_{stop} = 1$ s	$T_{stop} = 2$ s	$T_{stop} = 4$ s
Original CRN	0.05 millisec	167.78 s	339.42 s	676.03 s
Reduced CRN	0.05 millisec	46.12 s	94.5 s	128.88 s
Reduced CRN	0.5 millisec	4.67 s	9.64 s	18.5 s
Reduced CRN	1 millisec	2.57 s	4.7 s	9.46 s

which is registered to each time step of the monodomain equation solution. The reference profile has resting potential −80.83 mV. The Mitchell–Shaeffer model used in the monodomain equation solved by the lattice-Boltzmann algorithm uses the model parameters reported in [71], except for v_{gate}, which is set to 0.46. This corresponds to a voltage of −40 mV which is reported as the upstroke threshold for the I_{Na} channel in the original CRN model. We simulate the potential propagation on the whole anatomical model, with stimulus applied in the sino-atrial node. The temporal alignment of the regression cellular model is triggered when the transmembrane potential reaches $v = -10$ mV, after generation of the action potential. When repolarization is complete, and the transmembrane potential is $v < -75$ mV, the Mitchell–Schaeffer model is used to monitor the following upstroke.

Using the regression cellular model unlocks significant speedup in the solution of the monodomain problem. A direct comparison of the compute time required by the solution of the monodomain problem with CRN cellular model, versus the described set up, showed that using the regression cellular model reduces the computational cost by about 25% (see Table 4.6). For this comparison the time step dt is set to 0.05 milliseconds, which is sufficiently small to allow for a stable numerical solution of the CRN cellular model. When using the regression cellular model, the time step can be increased beyond this stability limit, since no numerical solution of the CRN model equation is required. In this scenario, a dramatic reduction of the computational cost is observed. For $dt = 1$ millisecond, the compute time was reduced to 1.5% of the time required for a full solution, and the simulation was near real-time. Examples of the reproduced temporal and spatial pattern of electrical potential in the considered atrial anatomical model are shown in Fig. 4.22.

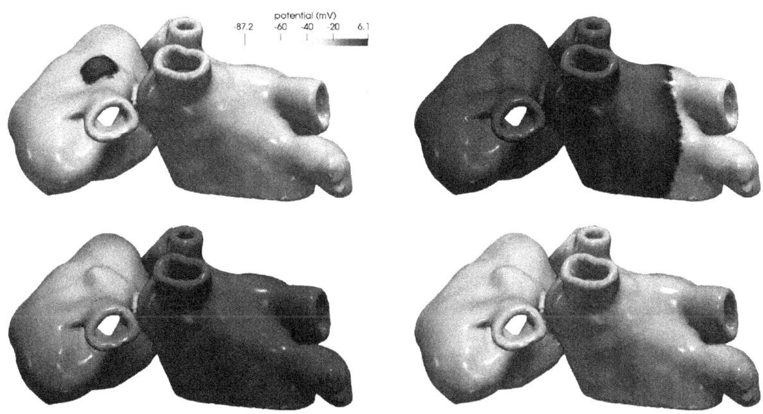

Figure 4.22. Simulated depolarization of a model of human atria, obtained by combining the regression-based model of action potential and the monodomain model solver. Top left: t =0 ms; Top right: $t = 100$ ms; Bottom left: $t = 200$ ms; Bottom right: $t = 300$ ms.

4.2.3 Discussion

This section presented methods to build high-fidelity electrophysiology cellular models using a data-driven approach. This enables efficient and accurate estimations of the action potential profile typical of complex, physics-based cellular models, by regression from the model parameters. The Courtemanche–Ramirez–Nattel (CRN) cellular model for human atria electrophysiology is used as an example of method application.

A simple way to integrate data-driven cellular model in the numerical solution of the monodomain model is also described, and demonstrated for the computation of the electrical activation of a realistic anatomical model of human atria. This integration approach relies on the use of a simplified, phenomenological cellular model for the explicit computation of the transmembrane potential dynamics, before and after the action potential. A natural extension of this work is the investigation of alternative methods for predicting upstroke, that do not rely on the explicit solution of a cellular model.

The methods presented can be applied to simulate more complex depolarization patterns, including the presence of different tissues, leading to different action potential morphologies. The extension to the modeling of abnormal rhythms is an important research direction, towards the application of this data-driven approach to whole heart electrophysiology modeling and simulation of pathological phenomena like arrhythmias, fibrillations or tachycardia.

4.3 Deep learning acceleration of biomechanics

4.3.1 Motivation

So far we have seen how data-driven techniques can be leveraged to drastically reduce the computational complexity in simulations of atrial electrophysiology and coronary hemodynamics. The simulation of cardiac contraction could also significantly benefit from such techniques. Biomechanics modeling of soft tissue consists in solving the deformation dynamics equation, as described in section 1.3. A common approach to numerically solve the underlying PDE is the finite element method (FEM). While producing highly accurate solutions, its computational complexity is high. Different approximation methods have been proposed, such as model order reduction or warping transformations, to reduce computational cost at the price of reduced accuracy. Providing accurate solutions in real-time for complex structures is still an open challenge. In this section, a deep learning based method is introduced to speed-up the simulation of soft tissue mechanics. The method relies on a machine trained model of motion able to describe various materials, geometries and load configurations. The Total Lagrangian Explicit Dynamics (TLED) algorithm from section 2.3.4 is used to generate training data and compute the input features, although any other FEM solver could also be used. The content of this section is based on the work presented in [369].

4.3.2 Methods

The core idea is to formulate the numerical solution of the time-dependent elastodynamics equation as a regression problem $\mathcal{R}$. Given the state of the system $\Phi(t)$ at time t, a neural network is trained to predict node-wise accelerations $\ddot{\mathbf{u}}_{\Delta t}(t)$ for a given time step Δt. Using a central-difference approximation for the ground-truth acceleration $\ddot{\mathbf{u}}_{\Delta t}(t) = 1/\Delta t^2[\mathbf{u}(t+\Delta t) - 2\mathbf{u}(t) + \mathbf{u}(t+\Delta t)]$ with the displacement $\mathbf{u}(\cdot)$ at a specific point in time, the new time integration formula writes $\mathbf{u}(t+\Delta t) = \ddot{\mathbf{u}}_{\Delta t}(t)\Delta t^2 + 2\mathbf{u}(t) - \mathbf{u}(t-\Delta t)$. This replaces traditional time-advancing numerical schemes, with the potential advantage of controlling the accuracy of the approximation even for large time steps (which cause instability of explicit time integration schemes and reduced accuracy of implicit schemes).

We now need to represent the state of the system. To this end, we rely on the instantaneous acceleration $\ddot{\mathbf{u}}_T(t)$, the backwards velocity $\dot{\mathbf{u}}_{\Delta t}(t) = [\mathbf{u}(t) - \mathbf{u}(t-\Delta t)]/\Delta t$, and the displacement $\mathbf{u}(t)$ at

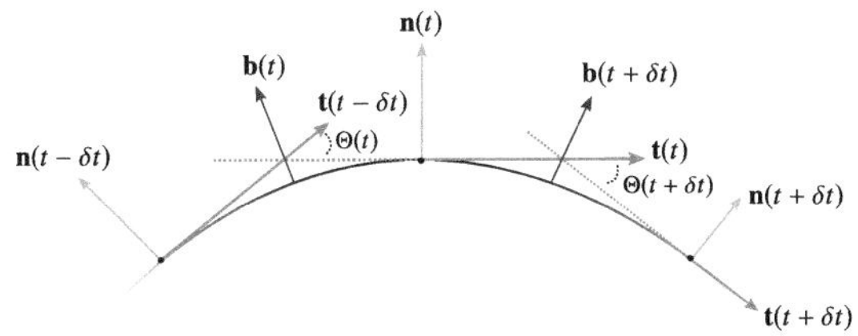

Figure 4.23. Visualization of the local coordinate system based on the parallel transport algorithm [370]. An initial coordinate system, defined by the tangent and two normal vectors, is iteratively rotated by the angle Θ between two subsequent tangent vectors. The rotation axis is defined by $\mathbf{b}(t) = \mathbf{t}(t) \times \mathbf{t}(t - \Delta t)$.

each node. The displacement is used for its translation-invariant representation of the vertex position with respect to the initial position. The velocity is used to account for inertia. The instantaneous acceleration is computed as $\ddot{\mathbf{u}}_T(t) = [\mathbf{f}_e - \mathbf{K}(\mathbf{u}(t))\mathbf{u}(t)]/m$, where m is the nodal mass. $\mathbf{f}_e$ denotes the external forces and $\mathbf{K}(\mathbf{u}(t))\mathbf{u}(t)$ are the internal material forces, which are computed using TLED. Thus, $\ddot{\mathbf{u}}_T(t)$ depends on the total force acting on each node of the computational domain.

While these features are by definition translation-invariant, they are not rotation-invariant. Consequently, it would require significant rotational augmentation of the training data to sample every possible motion. Instead, the features are expressed in a local coordinate system (see Fig. 4.23), which is defined by the tangent vector and two normals to the trajectory (Frenet–Serret frame). More precisely, the vector is aligned to the velocity and is computed as $\mathbf{t}(t) = \frac{\dot{\mathbf{u}}(t)}{\|\dot{\mathbf{u}}\|}$. The two normals are given by $\mathbf{n}_1(t) = \mathbf{R}(\mathbf{b}, \Theta) * \mathbf{b}(t - \Delta t)$ and $\mathbf{n}_2(t) = \mathbf{n}_1(t) \times \mathbf{t}(t)$. $\mathbf{R}$ denotes the rotation matrix defined by the angle $\Theta = \arccos(\mathbf{t}(t) \cdot \mathbf{t}(t - \Delta t))$ and rotation axis $\mathbf{b}(t) = \mathbf{t}(t) \times \mathbf{t}(t - \Delta t)$. This formulation allows the compression of the velocity feature to its magnitude.

The resulting feature vector is used as input to the network (Fig. 4.24). A grid search is applied to find the best number and dimension of the hidden layers. The final architecture consists of six fully-connected layers of decreasing size. The first five layers incorporate rectified linear units for the non-linear transformation, while the output layer uses the identity function to produce the acceleration prediction. A common problem of sequential prediction is the error accumulation, which could drive the features beyond what was observed during training. To ensure valid inputs to our model, the magnitude of each feature is scaled to be within the range of observed data during training.

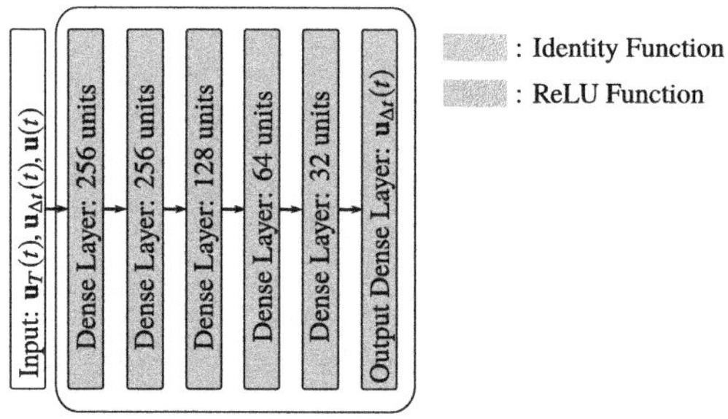

Figure 4.24. Illustration of the proposed fully-connected neural network architecture. Five non-linear layers and a linear output layer predict node-wise acceleration.

4.3.3 Evaluation

The approach was evaluated on a set of synthetic experiments comprising different geometries, motions, and materials. To this end, neural networks were trained for a specific time step. For training, the torsion of a hyperelastic bar modeled by the Holzapfel–Ogden material law was used. The bar measured $50\times50\times106$ mm^3. The mesh comprised 6,528 tetrahedra with its 1458 vertices placed on a regular grid. The torsion was induced by applying displacement constraints to the four corner points of one side, while fixing the other one in place. The displacement of the four corner points increased linearly to 50 mm over an interval of 1 s. The tissue parameters of the isotropic Holzapfel–Ogden equation (section 2.3.1) were set to a =0.059, b =8.023, $a_f = b_f = 0$ and d_1 =60. Furthermore, the damping coefficient was set to $\mu = 100$. TLED provided the solution at a time step of $\delta t = 3 \times 10^{-5}$ s. A total of 2000 frames were sampled from the interval $t = [0.3\text{ s } 1.0\text{ s}]$. To simplify the problem, the first 0.3 s were discarded due to high frequency oscillations arising from the transient effect of the initial condition. In addition, the corner points subject to boundary conditions were not included in the samples.

A network for a time advance of $\Delta t = 3 \times 10^{-4}$ s was trained on the torsion results. For parameter optimization, the Adam algorithm was used over 200 epochs at a learning rate of 1×10^{-4} [371]. The parameters β_1 and β_2 were set to 0.9 and 0.999, respectively. A batch size of 1,024 samples was used. Furthermore, a mean squared error loss function was used. The implementation of the network and of parameter optimization was performed using Keras with Tensorflow backend [373,374].

To test the networks ability to represent other motions and geometries, the model was applied to simulate the bending of a cylinder. The considered cylinder of size $50\times50\times101$ mm^3 con-

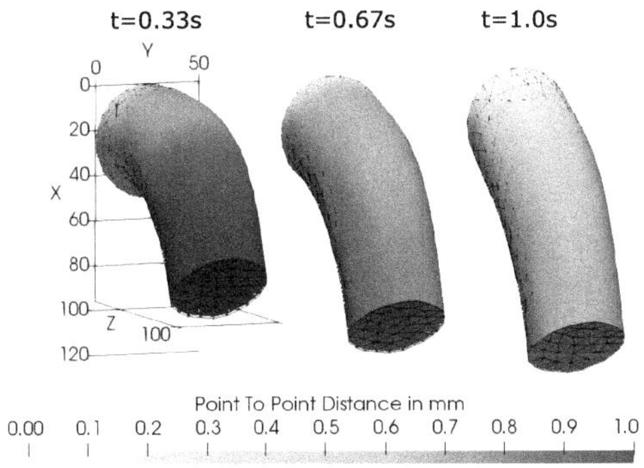

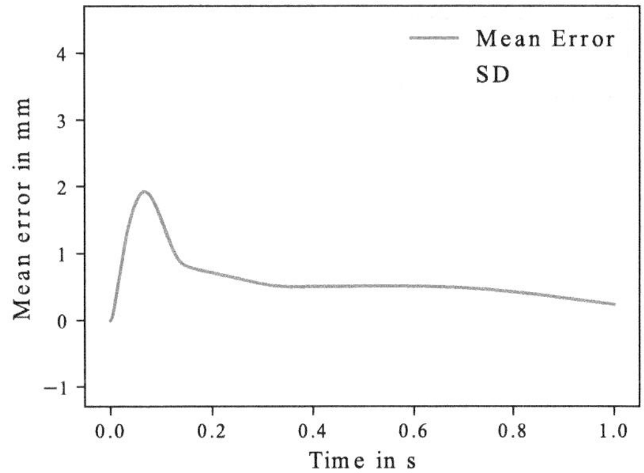

Figure 4.25. Illustration of the simulation results of cylinder bending modeled by the Holzapfel–Ogden model. The deep learning acceleration could simulate the deformation with an average error of 6.1×10^{-1} mm $\pm$ 8.2×10^{-1} mm over the entire simulation. In (A) the deformation at t=[0.33 s, 0.67 s, 1.0 s] can be seen. The meshes are color-coded by the point-wise error. In (B) the mean error over time can be seen.

sisted of 588 points and 2373 tetrahedra. The motion was induced by a normal stress of 0.5 MPa in positive x-direction at the plane z =101 mm. Furthermore, the plane at $z = 0$ mm was fixed in space. The simulation duration measured 1 s.

The network performed predictions at a time step ten times larger than used by the TLED solver to generate the training samples. In our experiments, TLED could not compute a stable solution for such a time step. In contrast, the time advancing scheme using deep learning based acceleration was capable of accurately simulating the motion with a mean error of 6.1×10^{-1} mm $\pm$ 8.2×10^{-1} mm. Fig. 4.25 shows the resulting deforming mesh color-coded by the point-wise error and a visualization of the mean error over time.

The application to other materials was also investigated. The same network was applied to simulate the compression of the bar, which was used during training. The material response was modeled by the Neo-Hookean material law [372], with strain-energy function $\psi = c(I_1 - 3) + d(J - 1)^2$ and tissue parameters set to c =0.5 and d =15. While TLED was not stable in this configuration, the deep learning method accurately simulated the deformation. Over the entire simulation a mean error of 0.6×10^{-3} mm $\pm$ 0.9×10^{-3} mm was observed. A visual comparison of the final deformation and the associated point-wise error can be seen in Fig. 4.26.

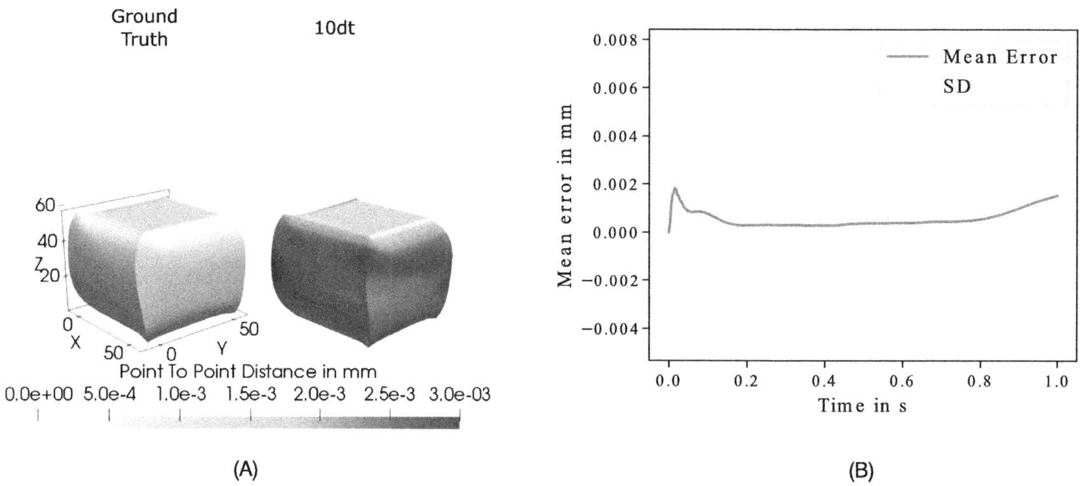

Figure 4.26. Illustration of the experimental results to recover the compression of a bar with a different material law than used during training. The deep learning method accurately simulated the compression with an average error of 0.6×10^{-3} mm $\pm$ 0.9×10^{-3} mm over time. In (A) a comparison between the final deformation computed using TLED and using the neural network is visualized. In (B) the mean error over time can be seen.

In addition, the method was evaluated to compute time advances with various time steps. To this end, networks were trained for a time step of 20dt, 30dt, 40dt, 50dt, 75dt, and 100dt and applied to simulate the cylinder bending. Up to 20dt our method demonstrated accurate simulations. For time steps beyond 20dt, high frequency oscillations and an artificial stiffening was observed. The error increased significantly. A comparison of the final deformation for all networks can be seen in Fig. 4.27.

4.3.3.1 Discussion

In this section an approach was introduced to accelerate biomechanical simulations, based on a machine trained motion model which does not require re-training if applied to other geometries, motions, and materials than used during training. While TLED was used to generate training data and to compute the features, the method is not bound to it and could be used in combination with other FEM solvers.

Given the current state of the system, the described method predicts point-wise accelerations for large time steps Δt, beyond the stability limit of TLED. Making point-wise predictions allows the application to various geometries as was illustrated by simulating the bending of a cylinder as opposed to the torsion of a bar, which was used during training. In addition, a local coordinate

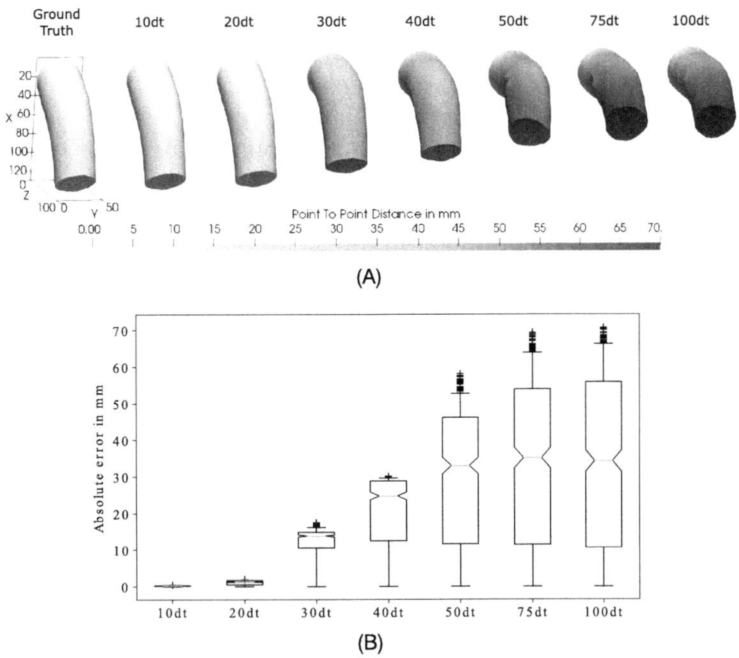

Figure 4.27. Visualization of the cylinder bending results for networks trained for different time steps (10dt, 20dt, 30dt, 40dt, 50dt, 75dt, and 100dt). In (A) the final deformation color-coded by the point-wise error is illustrated and the mean error can be seen in (B). While the network produced accurate results up to 20dt, the error increased for larger time steps due to an apparent artificial stiffening.

system was implemented, which reduces the number of training samples required to simulate arbitrary motions. Furthermore, in this framework the forces applied to the mesh vertices are used as input for the motion model. As a result, FEM solvers can be used to compute the forces at each time step, incorporating changes to material model and boundary conditions without having to retrain the AI model. The method was tested by simulating the compression of a bar using a Neo-Hookean material response instead of the Holzapfel–Ogden law. In all configurations TLED was unstable at a time step of 10dt. Through AI acceleration, conversely, excellent results with a sub-millimeter accuracy in retrieving the ground truth motion could be achieved.

While these results are promising, the geometries and motions used as test scenarios are simple and do not reflect real-world problems. Extension to more complex configurations will be subject of future work. Furthermore, although the approach was capable of accurately predicting the deformations up to a time step of 20dt, for large time steps an artificial stiffening and high-frequency oscillations were observed, which lead to large errors. Possible reasons for this effect could be insufficient training data and network capacity. Another reasons could be that the input features do not include explicitly information about the neighboring points, but rather only account for neighbors indirectly through the computation of internal forces. Explicitly adding features describing the neighborhood of each point could simplify model training in more complex scenarios. In addition, high errors might arise from the accumulation of the prediction error over a number of time steps. Although the magnitude of the input features is scaled to be within the range of observed data during training, error propagation could cause the input features to progressively drift towards values that are poorly sampled in the training database. A more detailed analysis will be the focus of future work.

4.4 Summary

This chapter presented three data-driven approaches to estimate surrogate models or accelerate numerical methods, with the goal to perform real-time hemodynamics, electrophysiology and biomechanics simulations. A deep learning model for FFR assessment was introduced, which yielded high accuracy in identifying functionally significant coronary artery lesions while by-passing time-consuming computational fluid dynamics simulations. An approach to estimate a meta-model of a complex cellular model, while preserving its physiological fidelity, has been introduced and demonstrated on a cardiac electrophysiology use case. Finally, a method to accelerate time integration in numerical solvers of biomechanics has been described. In all three cases, the combination of existing models, to create the training data in particular, and astute integration of neural networks in the simulation workflow allowed to achieve high computational efficiency while being as accurate as the original models. These examples are only a few of the many potential applications of the combination of artificial intelligence and computational modeling. Among other research areas that are opening in the same context, it is worth mentioning the integration of generative neural networks with computational models [375], and the estimation of new biophysical laws that can take into account data noise and uncertainty directly, for more accurate predictions [376].

5

Machine learning methods for robust parameter estimation

Dominik Neumann, Tommaso Mansi
Siemens Healthineers, Princeton, NJ, United States

5.1 Introduction

Individualizing computational models with patient data is a critical step for clinical applications. The goal consists in estimating the parameters $\boldsymbol{x}$ of a forward model f such that observable output variables of the model, denoted $\boldsymbol{y}$, match clinical observations $\boldsymbol{z}$ (Fig. 5.1). For instance, when the model f simulates cardiac electrophysiology (as described in section 1.2), the parameters $\boldsymbol{x}$ are the electrical conductivities, action potential durations, and other parameters associated to the cellular model of action potentials. The output $\boldsymbol{y}$ is the electrocardiograms (ECG) and related parameters, local activation times, etc., which can be measured using ECG or cardiac mapping devices. Contrary to the forward model, the inverse model infers the model parameters $\boldsymbol{x}$ from clinical measurements $\boldsymbol{z}$ [377].

A large variety of model personalization algorithms have been proposed in the last decades. Most methods follow a similar strategy: they iteratively try to find the parameter values $\hat{\boldsymbol{x}}$ that minimize the misfit between model output and measurements. In other words, estimating model parameters is traditionally posed as an optimization problem, where one seeks to optimize an objective c defined as the misfit between the computed $\boldsymbol{y}$ and measured $\boldsymbol{z}$ parameters: $\hat{\boldsymbol{x}} = \text{argmin}_{\boldsymbol{x}}\, c\,(\boldsymbol{y}, \boldsymbol{z})$. The estimation is considered successful if a convergence criterion is met, e.g. $|c| < \psi$ where ψ is a defined threshold. Different optimization techniques have been proposed, for instance variational techniques, trust region methods or highly engineered procedures [378,193,379].

Unfortunately, these techniques can suffer from large computational burden, low robustness and lack of accuracy, due to parameter ambiguity, data noise and local minima [201,26,82]. The need for a robust algorithm that works on larger populations [6, 195,380] often leads to more complex methods, which may combine numerous highly engineered optimizers. Typically, the de-

Artificial Intelligence for Computational Modeling of the Heart
https://doi.org/10.1016/B978-0-12-817594-1.00016-4

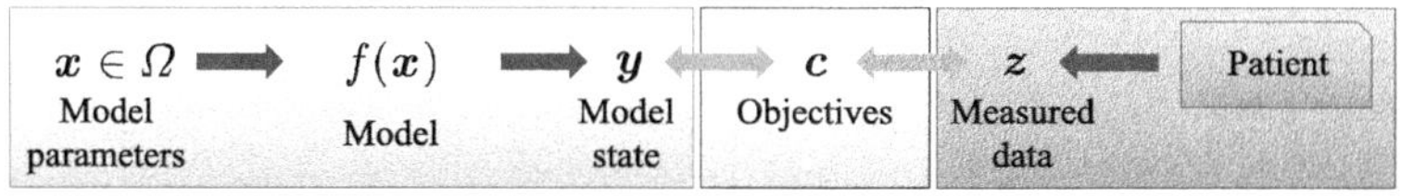

Figure 5.1. A computational model f is a dynamic system that maps model input parameters x to model state (output) variables y. The goal of personalization is to tune x such that the objectives c, defined as the misfit between y and the corresponding measured data z of a given patient, are optimized (the misfit is minimized).

sign of such methods is trial-and-error-driven and involves tedious manual tuning, which results in methods that are model-specific rather than generic, and still do not necessarily ensure generalization to unseen data.

Several machine learning techniques have been explored in the community to cope with the known limitations of optimization techniques. For instance, Jiang et al. [381] demonstrated a method to learn a mapping between body surface potentials maps and epicardial potentials, without requiring complex optimization procedures. Bayesian inference was also explored in [82] to estimate tissue conduction properties from invasive local activation time measurements. Finally, model reduction approaches as discussed in the previous chapter were applied to estimate a significantly faster surrogate model, which was then applied to estimate model parameters and their uncertainty [201]. These techniques demonstrated superior computational efficiency compared to traditional optimization approaches, and also the capability of providing more information about the estimated parameters (such as uncertainty).

On the other hand, when experienced humans are tasked with manually individualizing a model from patient data, they almost always succeed. There are several reasons why human experts are often superior to automatic methods. First, human experts often have the intuition of how a model behaves based on prior knowledge of the modeled physiology. Second, an expert usually knows the model design, assumptions, limitations and implementation details. Finally, experience built from past similar attempts on other datasets gives advantages. A combination of these and potentially other factors can enable a human expert to individualize models efficiently, even based on unseen data.

Motivated by the promising results of machine learning applied to parameter estimation, this chapter describes two approaches applied to concrete cardiac modeling use cases: a supervised learning method based on polynomial regression to directly estimate cardiac parameters from 12-lead ECG features, and a reinforcement learning method to train an agent that estimates model parameters using a learned strategy. Both methods reach either similar or superior performance compared to highly engineered inverse problem techniques, while being more efficient and precise.

5.2 A regression approach to model parameter estimation

A first data driven approach consists in learning a regression model to estimate parameters of a physiological model directly from clinical measurements, as previously described in [220]. Without loss of generality, the method is presented in the context of cardiac electrophysiology (EP), but it could be applied to models describing other aspects of cardiac function or other organs. Furthermore, we will also illustrate how uncertainty estimates can be calculated by leveraging the large database used to train the regression model.

5.2.1 Data-driven estimation of myocardial electrical diffusivity

The monodomain model introduced in section 1.2.2 is used without loss of generality as any other EP model could be also employed. The EP model $\boldsymbol{y} = f(\boldsymbol{x})$ takes as input the diffusivity coefficients $\boldsymbol{x} = (c_{Myo}, c_{LV}, c_{RV})$, among other parameters that are fixed in this example, and returns the ECG features $\boldsymbol{y} = (\text{QRS}_d, \alpha)$, where QRSd is the QRS complex duration measured on 12-lead ECG traces, and α is the electrical axis of the QRS complex. The goal is to learn a function g that approximates the inverse problem $\boldsymbol{x} = g(\boldsymbol{z}) \approx f^{-1}(\boldsymbol{z})$, namely that takes as input measured 12-lead ECG parameters ($\boldsymbol{z}$), and returns an estimate of the model parameters $\boldsymbol{x}$.

In this setting, though, ground truth data, namely pairs of clinical measurements ($\boldsymbol{z}$) and diffusivity values ($\boldsymbol{x}$), is not available. We therefore leverage the computational model to create a large database of pairs $(\boldsymbol{y}, \boldsymbol{x})$, assuming the model is accurate enough to provide realistic ECG features. The database is created by uniformly sampling the space of parameters $\boldsymbol{x}$, feeding the model with the sampled parameters and collecting the respective outputs $\boldsymbol{y}$. Once the database is ready, one could directly learn a regression model $\boldsymbol{x} = g(\boldsymbol{y})$. Unfortunately this task can be made difficult by the lack of observability of the parameters. Two sets of parameters $\boldsymbol{x}_1$ and $\boldsymbol{x}_2$ could give very similar observed ECG

Table 5.1 Diffusivity coefficients x in mm^2/s of the forward model evaluations used for normalization.

	c_{Myo}	c_{LV}	c_{RV}	Physiology
$\boldsymbol{x}_1$	100	4900	4900	Normal
$\boldsymbol{x}_2$	100	100	4900	LBBB
$\boldsymbol{x}_3$	100	4900	100	RBBB

parameters $\boldsymbol{y}_1 = f(\boldsymbol{x}_1) \approx \boldsymbol{y}_2 = f(\boldsymbol{x}_2)$. Furthermore, other determinants that are not part of the model parameters $\boldsymbol{x}$, in particular the heart shape and its position in the body, could contribute to having examples in the training database for which the same set of model parameters is associated to different ECG features. In other words, it is not guaranteed that the forward model is bijective. To cope with these limitations, several strategies can be used. For instance, the ECG features $\boldsymbol{y}$ can be normalized to the patient physiology. To that end, three forward evaluations with predefined diffusivity coefficients are done to scout the parameter space and estimate the range of variability of the ECG features (Table 5.1). These three "calibration" evaluations cover normal physiology (modeled by $\boldsymbol{x}_1$), left bundle branch block (LBBB) ($\boldsymbol{x}_2$) and right bundle branch block ($\boldsymbol{x}_3$) Given the three pairs $(\boldsymbol{y}_i, \boldsymbol{x}_i)_{i=1,2,3}$, the new ECG features are normalized as follows:

- The QRS duration is normalized using QRSd_1 which corresponds to normal cardiac activation: $\overline{\text{QRSd}} = \text{QRSd}/\text{QRSd}_1$
- The electrical axis is normalized given the range between $\boldsymbol{y}_2$ and $\boldsymbol{y}_3$: $\overline{\alpha} = (\alpha - \alpha_2)/(\alpha_3 - \alpha_2)$. This corresponds to assuming that the electrical axis will reach extremal values in left bundle branch block and right bundle branch block configurations.

We refer to the resulting normalized ECG features as $\overline{\boldsymbol{y}}$. Several regression approaches can be used to learn the function $\boldsymbol{x} = g(\overline{\boldsymbol{y}})$. We here consider multivariate polynomial regression of degree N, but other approaches like multivariate adaptive regression splines (MARS) and Gradient Boosting [382] can be used. To simplify the learning, the parameters are estimated component wise. In particular, for polynomial regression, one function of the form

$$g(\overline{\text{QRSd}}, \overline{\alpha}) = \sum_{i=0}^{N} \sum_{j=0}^{N} \beta_{i,j} \left(\overline{\text{QRSd}}\right)^i (\overline{\alpha})^j + \varepsilon \tag{5.1}$$

is learned for each diffusivity parameter independently, $\boldsymbol{x} = (c_{Myo}, c_{LV}, c_{RV})$. The training procedure consists in estimating the coefficients $\beta_{i,j}$. During testing, i.e. when one wants to estimate the diffusivity coefficients from ECG data on an unseen case, the ECG parameters are first normalized and then the function g is applied.

5.2.2 Experiments and results

5.2.2.1 Setup and uncertainty analysis

The method presented in this section was comprehensively evaluated in [220]. We report in the following the main highlights. The evaluation was performed on a dataset of 19 patients with dilated cardiomyopathy (DCM) and a QRS duration higher than 120 ms. Patients were eligible for CRT according to the current guidelines of the European Society of Cardiology and received treatment at University Hospital Heidelberg. The workflow described in chapter 2 (Fig. 2.1) was followed to create the anatomical model from magnetic resonance images (MRI). For each case, 500 different EP simulations were performed, making a total of 9500 different simulations. Cardiac EP was calculated using the Lattice-Boltzmann solver for the monodomain problem (section 2.2.1) using the Michell-Schaeffer cellular model [71], with an isotropic grid resolution of 1.5 mm. The diffusivity coefficients were randomly sampled according to a uniform distribution between 50 mm^2/s and 5000 mm^2/s, with $c_{Myo} \leq c_{LV}, c_{RV}$.

Leveraging the large number of simulations, the uncertainty of the inverse problem could be estimated. In other words, one could assess the spread of diffusivity values for which the forward model would produce very similar values of QRS duration and electrical axis. To factor out the effect of the heart geometry and position, the analysis was done on the normalized ECG parameters. The pairs $(\boldsymbol{x}, \boldsymbol{y})$ were clustered in 20×20 bins along the $\overline{QRSd}$ and $\overline{\alpha}$ axes respectively. The bin-wise standard deviations of c_{Myo}, c_{LV} and c_{RV} were calculated. The uncertainty of the diffusivity coefficients was then defined as the ratio between the average of all bin-wise standard deviations over the total standard deviation (Table 5.2). Fig. 5.2 illustrates the bin-wise standard deviations. As one can see, the largest uncertainty happened in the healthy ranges for both QRS duration and electrical axis (up to 180%), confirming the ill-posed nature of the EP inverse problem.

5.2.2.2 Verification on synthetic data

A first evaluation consists in verifying whether the learned regression model is able to recover the true diffusivity coefficients on the simulated database, using a leave-one-patient-out strategy. A case was selected randomly for evaluation, while the re-

Table 5.2 Uncertainty of the diffusivity coefficients when estimated from QRS duration and electrical axis, using LBM-EP.

	c_{Myo}	c_{LV}	c_{RV}
Total SD (σ_T, mm^2/s)	1482	1095	1191
Avg. bin-wise SD ($\overline{\sigma_i}$, mm^2/s)	191	556	537
Uncertainty ($\overline{\sigma_i}/\sigma_T$)	12.9%	50.7%	45.1%

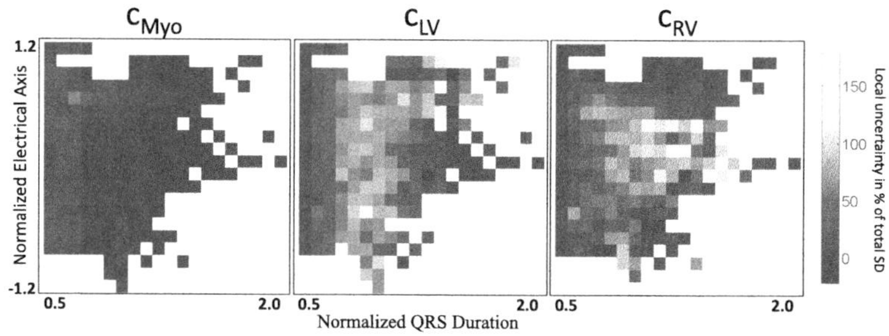

Figure 5.2. Bin-wise standard deviation (SD) in % of total SD of each diffusivity coefficient, for known electrical axis and QRS duration. The regions of highest uncertainty are clustered at the center of the plots, i.e. within the healthy range of each parameter.

maining 18 cases were used to train the regression model. Using the simulated data allowed to get a direct, quantitative evaluation of the regression model by averaging the absolute difference between the actual diffusivity coefficients (which was used to get the QRS duration and electrical axis features) and the estimated ones. We could also evaluate the goodness of fit, namely the difference between the ground truth QRS duration and electrical axis, and the values computed using the model and the diffusivity parameters estimated by the regression model.

To establish the optimal degree of the polynomial regression, a cross-validation was performed within the training set. Fig. 5.3 illustrates the performance obtained at degrees from 1 to 8. From this analysis, one could assess that low-degree polynomials did not perform well, likely due to their inability to approximate the solution manifold of the inverse problem. On the other hand, considering polynomials of degree 5 and beyond also did not bring performance benefits, likely due to overfitting of the training sam-

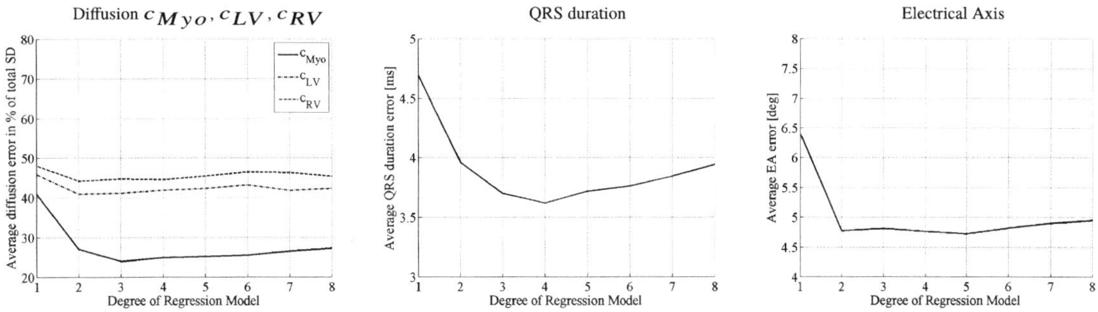

Figure 5.3. Prediction accuracy of polynomial regression models with increasing degrees. The optimal compromise between performance and over-fitting was achieved with polynomials of degree 3 or 4.

Table 5.3 Error in parameter estimation, in absolute values (mm^2/s) and in % of the total standard deviation. The regression model could estimate diffusivity values up to the inherent uncertainty of the problem (Table 5.2). Normalizing the input features significantly improved the performances of the estimation algorithm.

	c_{Myo}	c_{LV}	c_{RV}
With normalization	356	451	533
	24.0%	41.2%	44.7%
Without normalization	571	540	597
	38.5%	49.3%	50.0%

ples. Polynomials of degree 3 were therefore considered for the final model, and used in all the subsequent experiments.

Table 5.3 reports the absolute differences between ground truth and predicted diffusivity values. The experiment showed that the regression model was able to provide estimates within the inherent uncertainty of the inverse problem, as reported in Table 5.2. To improve the accuracy, adding more data (e.g. parameterizing the output of the forward model by the full ECG trace rather than by a few handcrafted features) may therefore be more helpful than fine tuning the approach.

The effect of the normalization of the ECG parameters could also be verified (Table 5.3). When learning a mapping from the raw ECG parameters to the diffusivity coefficients directly, the error in estimated parameters was significantly higher. This result suggested an important effect of the heart geometry and position in the torso, which could be compensated, at least partially, by the proposed normalization procedure.

Table 5.4 Mean error in ECG features using diffusivity parameters regressed from clinical measurements.

Diffusivity	$QRSd$ [ms]	α [deg]
Regression-based prediction	18.7 ± 16.2	6.5 ± 7.6

5.2.2.3 Evaluation on patient data

The method was then evaluated on the 19 patient data. Because ground truth diffusivity coefficients cannot be measured directly in patients, we evaluated the goodness of fit of the personalization approach, namely how close were the computed QRS duration and electrical axis after personalization from the measurement. The regression model trained in the previous section was used for the personalization. Table 5.4 reports the obtained results. The personalization failed in 3 cases ($\approx$ 16% of the cases), yielding negative diffusivity coefficients. The reason was that the measured electrical axis for these cases was outside of the normalization range used for training, potentially due to an atypical position of the heart within the torso. Such a situation could easily be detected and other approaches could then be used to estimate the diffusivity parameters. The diffusivity coefficients estimated for the other patients were within expected ranges ($c_{Myo} \in [141, 582]$ mm^2/s, $c_{LV}, c_{RV} \in [678, 2769]$ mm^2/s). As reported in Table 5.4, the final error between measured and computed QRS duration was about 18 ms and between the measured and computed electrical axis was equal to 6.5°, both accepted clinically as within the noise in the measurement. Finally, Fig. 5.4 reports the overlay of simulated ECG leads on measured traces for one representative patient, showing promising goodness of fit.

5.3 Reinforcement learning method for model parameter estimation

The method described in the previous sections was based on traditional supervised learning. What if there was an algorithm that can learn by itself how to personalize a model? To this end, we reformulate the problem in terms of reinforcement learning (RL) [266]. RL has its roots in control theory and neuroscience theories of learning. It encompasses a set of approaches that make an artificial agent learn from experience generated by interacting with its environment. Contrary to supervised learning [331],

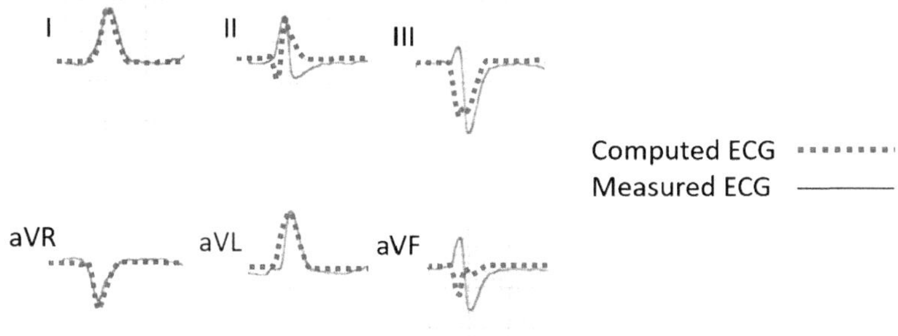

Figure 5.4. Measured and computed ECG traces for one representative cases (estimation errors of 1.6 ms for QRS duration and 0.5° for electrical axis).

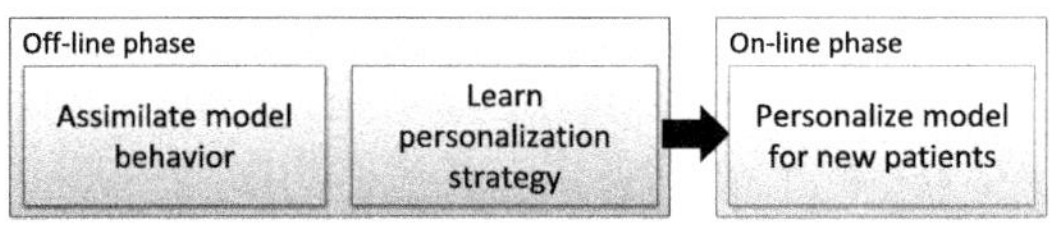

Figure 5.5. Framework overview: self-taught artificial model personalization agent.

where the objective is to compute a direct mapping from a given input to a prediction, RL aims to learn how to perform tasks. This is achieved by computing an optimal strategy, called "policy", to solve a given problem. A policy describes a mapping from states, i.e., the current "situation" the agent finds itself in, to actions, which allow the agent to interact with the environment. Rewards allow the agent to judge the outcome of its actions.

The past few years saw tremendous breakthroughs in RL for complex, real-world problems, with application to medical imaging [383]. One instance is presented in chapter 3, where an RL agent learns how to detect landmarks in 3D medical images, while being faster, more accurate and more generic than alternative algorithms.

Motivated by these recent successes, we proposed an RL-based personalization approach called Vito [384], a class of artificial agents that learn by themselves how to estimate model parameters from clinical data while being model-independent. First, in an off-line data-driven exploration phase, Vito assimilates the behavior of physiological models (Fig. 5.5). Based on the gathered knowledge, Vito then learns an optimal personalization strategy using RL. Once the off-line phase has converged, given a new, unseen dataset, Vito sequentially chooses actions that maximize future rewards, i.e., that bring the agent into a state that represents the solution of the personalization problem. To set up the algorithm, the user has to define the observations to be matched, the available actions, and a single hyper-parameter related to the state space configuration. Then everything is learned automatically. The framework does not depend on the biophysical model to be personalized. We evaluated it on two different models: the inverse problem of cardiac electrophysiology and the personalization of a lumped parameter model of whole-body circulation.

5.3.1 Parameter estimation as a Markov decision process

To apply RL to a problem, we first need to map it into a Markov Decision Process (MDP) [266]. In brief, an MDP is defined as a tuple $\mathcal{M} = (\mathcal{S}, \mathcal{A}, \mathcal{T}, \mathcal{R}, \gamma)$, where $\mathcal{S}$ is the finite set of states describing the environment, $\mathcal{A}$ is the finite set of actions for interacting with the environment, $\mathcal{T}$ is the stochastic transition function, where $\mathcal{T}(s_t, a_t, s_{t+1})$ describes the probability of arriving in state s_{t+1} after the agent performed action a_t in state s_t, $\mathcal{R}$ is the reward function, where $r_{t+1} = \mathcal{R}(s_t, a_t, s_{t+1})$ is the immediate reward the agent receives after performing action a_t in state s_t resulting in state s_{t+1}, and $\gamma \in [0; 1]$ is the discount factor. The goal of RL is to find the optimal policy $\pi^* : \mathcal{S} \rightarrow \mathcal{A}$, i.e., the mapping from states to actions that maximizes the expected value of the cumulative discounted reward. The optimal policy for a fully defined MDP can be found by applying the value iteration method [266], among other techniques.

However, since not all MDP components are known precisely ($\mathcal{T}$ is only an approximation from training data as we will see later), value iteration does not guarantee optimality. To mitigate potential issues due to this, we use a stochastic policy $\tilde{\pi}^*$ [384], instead of the standard deterministic policy. For a given state, while a deterministic policy always returns the action with the highest state-action value (function computed by value iteration), the stochastic policy keeps multiple candidate actions with similar high state-action value (threshold defined by user), and returns one of them through a random process each time it is queried.

5.3.1.1 Reformulation of model personalization into an MDP

The model personalization problem is mapped to an MDP as follows:

- **States** encode the misfit between computed model output and patient's measurements. While misfit is generally continuous, the number of MDP states has to be finite, therefore the space of objective vectors, $\mathbb{R}^{n_c}$, is reduced to a finite set of representative MDP states $\mathcal{S}$, each $s \in \mathcal{S}$ covering a small region of that space. $\hat{s} \in \mathcal{S}$ denotes the success state, which covers the region where all convergence criteria are met. The overall goal is to learn how to reach that state.

- **Actions** modify the parameters $\boldsymbol{x}$ to fulfill the objectives $\boldsymbol{c}$. An action $a \in \mathcal{A}$ consists in either in- or decrementing one parameter x_i by $1\times$, $10\times$ or $100\times$ a user-specified reference value $\delta_i \in \boldsymbol{\delta}$. This quantization of the intrinsically continuous action space can be defined empirically.
- **Transition function** encodes learnt knowledge about the computational model f, see following paragraphs for details.
- **Rewards** are defined as $\mathcal{R}(s, a, s') = -1$ (punishment), except when an action resulting in personalization success was performed, then $\mathcal{R}(\cdot, \cdot, \hat{s}) = 0$.
- **Discount factor** is designed to encourage finding a policy favoring future over immediate rewards to avoid local minima, therefore a large $\gamma = 0.99$ is used.

5.3.1.2 Learning model behavior through exploration

To learn the behavior of the model f when its parameter values are changed, we perform model exploration episodes. Each episode is initiated with random, but physiologically plausible model parameters $\boldsymbol{x}_t \in \Omega$ and one of the available training patients. From the outputs of a forward model run $\boldsymbol{y}_t = f(\boldsymbol{x}_t)$, the misfits to the patient's corresponding measurements $\boldsymbol{z}$ are computed, yielding the objectives vector $\boldsymbol{c}_t = c(\boldsymbol{y}_t, \boldsymbol{z})$. We then employ a random exploration policy π_{rand}, i.e., a policy that selects random actions according to a discrete uniform probability distribution. The selected action is applied to the current parameter vector, yielding $\boldsymbol{x}_{t+1} = a_t(\boldsymbol{x}_t)$. The forward model is computed, $\boldsymbol{y}_{t+1} = f(\boldsymbol{x}_{t+1})$, yielding $\boldsymbol{c}_{t+1}$. The next action a_{t+1} is selected according to π_{rand}, and so on. This process is repeated $n_{e\text{-steps}} = 100$ times in our experiments to complete one episode, which can be written as:

$$\{(\boldsymbol{x}_t, \boldsymbol{y}_t, \boldsymbol{c}_t, a_t, \boldsymbol{x}_{t+1}, \boldsymbol{y}_{t+1}, \boldsymbol{c}_{t+1}), \quad t = 0, \ldots, n_{e\text{-steps}} - 1\} \ . \tag{5.2}$$

Many of such episodes are usually generated per available training patient and later combined into one multi-patient set of episodes, denoted $\mathcal{E}$. It is important to have multiple training patients to allow abstraction from patient-specific to model-specific knowledge.

Please note this is done once, at training, and is not repeated for each personalization procedure. Exploration can therefore be as extensive as needed.

5.3.1.3 From computed objectives to representative MDP state

The continuous space of objective vectors now needs to be quantized into a finite set of representative MDP states $\mathcal{S}$ using a data-driven approach. To this end, all objective vectors $\boldsymbol{c}$ that were observed during exploration (as part of episodes in $\mathcal{E}$) are grouped into $n_{\mathcal{S}} - 1$ clusters based on their distance to each other. The distance metric is defined relative to the inverse of the thresholds in the convergence criteria to ensure similar influence of all objectives (e.g., to cancel out different units, etc.):

$$\|\boldsymbol{c}_1 - \boldsymbol{c}_2\|_{\boldsymbol{\psi}} = \sqrt{(\boldsymbol{c}_1 - \boldsymbol{c}_2)^\top \operatorname{diag}(\boldsymbol{\psi})^{-1} (\boldsymbol{c}_1 - \boldsymbol{c}_2)} \ . \tag{5.3}$$

The centroid of each cluster becomes the centroid of a representative state, and the special "success state" mentioned earlier, denoted $\hat{s}$, is artificially created to cover the region in state space where all objectives are met: $\forall i : |c_i| < \psi_i$. This results in a total of $n_{\mathcal{S}}$ states: $n_{\mathcal{S}} - 1$ are data-driven, and one is the success state.

To determine the MDP state of a given objective vector $\boldsymbol{c}$, we introduce a mapping ϕ. Let $\boldsymbol{\xi}_s$ denote the centroid corresponding to state s, then the mapping is defined as:

$$\phi(\boldsymbol{c}) = \underset{s \in \mathcal{S}}{\operatorname{argmin}} \|\boldsymbol{c} - \boldsymbol{\xi}_s\|_{\boldsymbol{\psi}} \ . \tag{5.4}$$

5.3.1.4 Transition function as probabilistic model representation

In this work, the stochastic MDP transition function $\mathcal{T}$ is generated such that the transition probabilities encode the learnt knowledge about the behavior of computational model f. To this end, we rely on model exploration and resulting training episodes $\mathcal{E}$. First, the individual samples $(\boldsymbol{x}_t, \boldsymbol{y}_t, \boldsymbol{c}_t, a_t, \boldsymbol{x}_{t+1}, \boldsymbol{y}_{t+1}, \boldsymbol{c}_{t+1})$ are converted to state-action-state transition tuples $\hat{\mathcal{E}} = \{(s, a, s')\}$, where $s = \phi(\boldsymbol{c}_t)$, $a = a_t$ and $s' = \phi(\boldsymbol{c}_{t+1})$. The transition function is then approximated at each point based on statistical analysis of the observed transition samples:

$$\mathcal{T}(s, a, s') = \frac{|\{(s, a, s') \in \hat{\mathcal{E}}\}|}{\sum_{s'' \in \mathcal{S}} |\{(s, a, s'') \in \hat{\mathcal{E}}\}|} \ . \tag{5.5}$$

Some state-action combinations may not be observed, especially if $n_{\mathcal{S}}$ and $n_{\mathcal{A}}$ are large. In such cases, uniform probability is assigned.

Now that the MDP model $\mathcal{M}$ is fully defined, we apply value iteration (section 5.3.1) and compute the stochastic policy $\tilde{\pi}^*_\epsilon$, which completes the off-line phase.

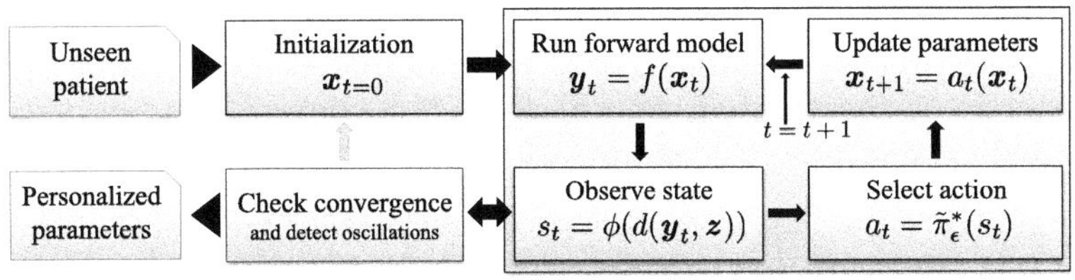

Figure 5.6. Probabilistic on-line personalization phase.

5.3.2 Parameter estimation using Reinforcement Learning

On-line personalization of unseen patients, as illustrated in Fig. 5.6, can be seen as a two-step procedure. First, we apply a data-driven initialization step to determine an initial guess for the parameters to be personalized. Second, we rely on the computed stochastic policy $\tilde{\pi}^*_\epsilon$ to perform the personalization.

5.3.2.1 Data-driven initialization

To initialize the personalization procedure, we rely again on the knowledge obtained during exploration, stored in $\mathcal{E}$. First, we look for episode steps, where the resulting model state $\boldsymbol{y}$ is similar to the current patient's measurements $\boldsymbol{z}$. Then we identify the input parameters $\boldsymbol{x}$ that led to these results: $\{\boldsymbol{x} \in \mathcal{E} \mid f(\boldsymbol{x}) \approx \boldsymbol{z}\}$, and analyze their distribution, which could be multi-modal due to ambiguities induced by the different training patients, noise in the data, modeling assumptions, etc. In particular, we cluster the identified parameter vectors according to their similarity, compute one representative per cluster, and rank the clusters based on their size in descending order. The cluster representatives serve as initialization candidates.

5.3.2.2 Probabilistic personalization

As illustrated in Fig. 5.6, from the most likely initial parameters $\boldsymbol{x}_0 \in \{\boldsymbol{x}_0\}$, the forward model $\boldsymbol{y}_0 = f(\boldsymbol{x}_0)$ is run and the misfit between the model output and the patient's measurements $\boldsymbol{c}_0 = c(\boldsymbol{y}_0, \boldsymbol{z})$ is computed to derive the first state $s_0 = \phi(\boldsymbol{c}_0)$. Given s_0, the policy is queried to decide the first action to take $a_0 = \tilde{\pi}^*_\epsilon(s_0)$. The process runs through state-action-state sequences to personalize the computational model f by iteratively updating the model parameters through MDP actions. Bad initialization could lead to oscillations between states as reported in literature [385,386], which can be detected by monitoring the parameter traces to detect recurring sets of parameter values. If that happens, the personalization is re-initialized at the second-most-likely $\boldsymbol{x}_0 \in \{\boldsymbol{x}_0\}$, etc. The process terminates once the designated success state $\hat{s}$ is reached (success), or after a pre-defined maximum number of iterations (failure).

5.3.3 Application to cardiac electrophysiology

We tested the approach on cardiac electrophysiology for 83 dilated cardiomyopathy patients who received treatment at University Hospital Heidelberg. For each patient, the end-diastolic biventricular anatomy was segmented from short-axis cine MRI, a tetrahedral anatomical model including myofibers was estimated and a torso atlas registered to the patient. Please refer to section 2.1 for more details.

We consider a graph-based model of EP (see section 2.2.3), which is controlled by conduction velocities (in mm/s) of myocardial tissue and left and right Purkinje network: $\boldsymbol{x} = (c_{\text{Myo}}, c_{\text{LV}}, c_{\text{RV}})^\top$. In the following experiments, the admissible parameter space Ω was set to [200; 1000] for c_{Myo} and [500; 5000] for both c_{LV} and c_{RV}. Reference increment values to build the action set $\mathcal{A}$ were set to $\boldsymbol{\delta} = (5, 5, 5)^\top$ mm/s for the three model parameters. The goal of personalization was to estimate $\boldsymbol{x}$ from the measured QRS duration and electrical axis (EA). Personalization was considered successful if the misfits in QRSd and EA were below $\boldsymbol{\psi} = (5\,\text{ms}, 10°)$, respectively.

Number of representative states Quantization of the state space relies on a single hyper-parameter $n_{\mathcal{S}}$, the number of representative states (see section 5.3.1.3). We used eight scouting patients to find a good value for $n_{\mathcal{S}}$ using exhaustive search for $n_{\mathcal{S}} \in \{10, 20, \ldots, 490, 500\}$ representative states. The configuration quality was evaluated based on average number of required forward model runs and personalization success rate on the scouting patients. Good quality was observed between 50 to 300 representative states, with a peak at 120, where success rate was 97%, and where 101 forwards model runs were required on average. Therefore, $n_{\mathcal{S}} = 120$ was selected for further experimentation. The eight scouting datasets were discarded for the following experiments to avoid bias in the analysis.

Reference methods We compared the results of the RL-based method against two optimization approaches, both based on an advanced, gradient-free optimization technique called BOBYQA [187]:

- SIMPLEGEP mimicks the most basic estimation setup, where only the minimum level of model and problem knowledge were assumed. To this end, the sum of absolute QRSd and EA errors

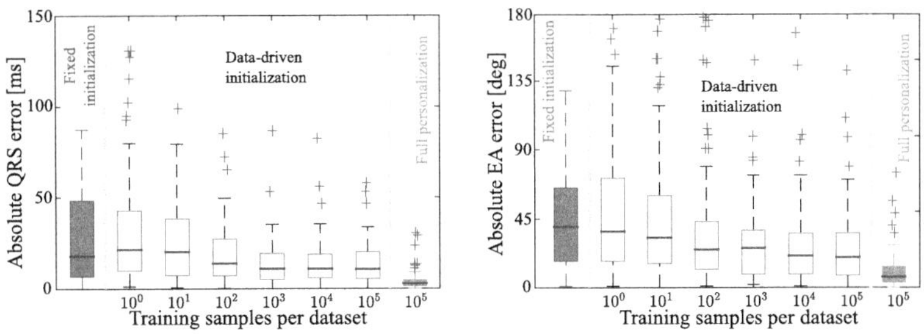

Figure 5.7. Absolute errors for all patients after initialization with fixed parameter values (blue, dark gray in print version), after data-driven initialization for increasing amount of training data (white), and after full personalization (green, light gray in print version). Data-driven initialization yielded significantly reduced errors if sufficient training data were available ($> 10^2$) compared to initialization with fixed values. Full personalization further reduced the errors significantly. Red (mid gray in print version) bar and box edges indicate median absolute error, and 25 and 75 percentiles, respectively.

were minimized by tuning all three parameters in $\boldsymbol{x}$ simultaneously. The algorithm terminated once all convergence criteria were satisfied (success) or if the number of forward model evaluations exceeded 100 (failure).

- CASCADEGEP implements an advanced estimator with strong focus on robustness, which computes the optimum parameters in a multi-step iterative fashion as described in [380].

To remove bias towards the choice of initial parameter values, for each of the two methods all datasets were personalized 100 times with different random, but physiologically plausible (Ω) initializations. We observed significant differences in performance with different initializations, indicating non-convexity of the cost function: the number of successfully personalized cases varied from 13 to 37 for SIMPLEGEP, and from 31 to 51 for CASCADEGEP, which means a variability of more than 25% of the total number of patients.

Full personalization performance First, we evaluated the overall performance of the RL-based method and compared against the reference methods. We ran the full personalization pipeline (off-line learning, initialization, and on-line personalization) on all patients with leave-one-patient-out cross-validation using 1000 training episodes per patient and maximum number of iterations set to 100. The green (light gray in print version) box plots in the two panels of Fig. 5.7 summarize the results. The mean absolute errors were 4.1 ± 5.6 ms for QRSd and $12.4 \pm 13.3°$ for EA. In comparison, the best SIMPLEGEP run yielded absolute errors of 4.4 ± 10.8 ms for QRSd and $15.5 \pm 18.6°$ for EA on average, and the best CASCADEGEP run 0.1 ± 0.2 ms for QRSd and 11.2 ± 15.8° for EA, respectively. In summary, all three methods yielded comparable performance in terms of EA error, SIMPLEGEP and the proposed method performed similarly in terms of QRSd and were outperformed in this regard by CASCADEGEP. However, considering success rates (successfully personalized patients according to the defined convergence criteria), both the performance of Vito (67%) and CASCADEGEP (68%) were equivalent, while SIMPLEGEP reached only 49% or less. In terms of run-time, i.e. average number of forward model runs until convergence, Vito (31.8) almost reached the high efficiency of SIMPLEGEP (best: 20.1 iterations) and clearly outperformed CASCADEGEP (best: 86.6 iterations), which means Vito was $\approx 2.5\times$ faster.

Residual error after initialization One major advantage over standard methods is the data-driven initialization step (see section 5.3.2.1), which eliminates the need for the user to provide initial parameter values. To evaluate the utility of this step, we evaluated the forward model with the computed initialization without further personalization, then quantified the resulting errors and compared them against the error after initialization when fixed initial values were used (based on initialization of the best performing BOBYQA run). This was done for increasing number of transition samples per dataset: $n_{\text{samples}} = 10^0 \ldots 10^5$. Fig. 5.7 shows that by increasing training data, both errors decreased notably. In fact, only 100 transitions per dataset suffice to become more accurate than the best tested fixed initial values.

In summary, the proposed initialization not only simplifies the setup, as it removes the need for user-provided initialization, this experiment also showed that it can reduce initial errors by a large margin with only few training transitions available. It should be noted again that in its normal operating mode (continue personalization after initialization), the model fit is further improved, as demonstrated in the previous experiment.

Convergence analysis With the next experiment, we investigated how much training data (transition samples) were needed to achieve solid performance of the agent. To this end, we evaluated the proposed method with varying number of training transition samples per dataset and found increasing performance with increasing training data (Fig. 5.8), suggesting that the learning process was working properly. At $n_{\text{samples}} = 10^2$ samples per patient, we already outperformed the best version of SIMPLEGEP (49% success rate). Starting from $n_{\text{samples}} \approx 3000$, a plateau at $\approx 66\%$ success rate was reached, which then remained approximately constant and almost on par with the top CASCADEGEP performance (68% success rate). Also the number of model runs until convergence

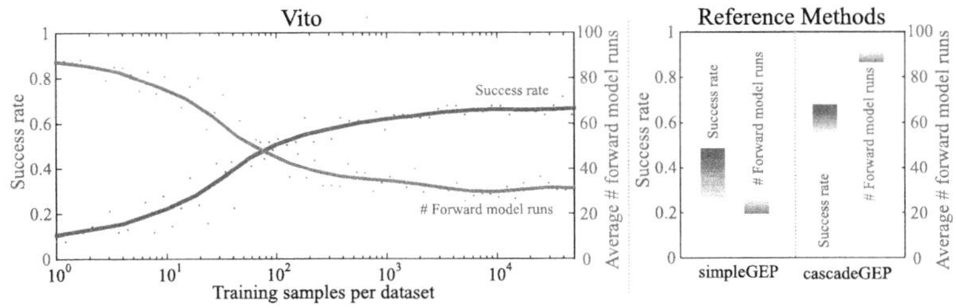

Figure 5.8. EP results: personalization success rate (blue, dark gray in print version) and average number of iterations (red, mid gray in print version). Left: performance for increasing number of training data. Each dot represents results from one experiment (cross-validated personalization of all 75 datasets), solid lines are low-pass filtered means. Right: Performance of both reference methods. Each shade represents 10% of the results, sorted by performance.

decreased when there was more training data available: 36.2 at 10^3 samples, 31.5 at 10^4 samples, or 31.8 at 10^5 samples.

These results suggested that the proposed method can achieve similar performance as an advanced, manually engineered method. Further, the number of required training samples was not excessive, instead well manageable and computable in a reasonable time-frame.

5.3.4 Application to whole-body circulation

The second considered task is the personalization of a lumped-parameter whole-body circulation (WBC) model from pressure catheterization and volume data. As some of the datasets from the EP experiments lacked catheterization data, only a subset of 56 patients was used in the following experiments. The WBC model is described in section 1.4.2 and [387,386]. In short, it consists in time-varying elastance models for the four chambers of the heart, the valves are modeled through a resistance and an inertance, and Windkessel (WK) models are used for the systemic and pulmonary circulation: 3-element WK for arterial and 2-element WK for venous circulation, respectively.

The goal of this experiment was to analyze personalization performance of the RL-based method for the part of the model describing systemic circulation, in various configurations. In particular, we evaluated the method for increasing number of parameters to tune and objectives to match, from two to six parameters (setups 2p, 3p, 5p, 6p), see Table 5.5 for details. The reference values δ that define the action space $\mathcal{A}$ were set to 0.5% of the admissible parameter range Ω. Personalization objectives are listed in Table 5.6 and illustrated in Fig. 5.9.

Table 5.5 WBC parameters x, their default values and domain Ω. The last column denotes the experiment setups in which a parameter was personalized. Default values were used in experiments where the respective parameters were not personalized.

x	Default value	Ω	*Setups*
Initial volume	400 mL	[200; 1000] mL	6, 5, 3, 2
Maximum elastance	2.4 mmHg/mL	[0.2; 5] mmHg/mL	6, 5, 3, 2
Aortic resistance	1100 g / (cm^4 s)	[500; 2500] g / (cm^4 s)	6, 5, 3
Aortic compliance	$1.4 \cdot 10^9$ cm^4 s^2/ g	$[0.5; 6] \cdot 10^9$ cm^4 s^2/ g	6, 5
LV dead volume	10 mL	[−50; 500] mL	6, 5
Time to E_{max}	300 ms	[100; 600] ms	6

Table 5.6 WBC output, the threshold ψ in the corresponding convergence criteria and range of measured values in the patient population used for experimentation.

Model output	ψ	Measured range	*Setups*
End-diastolic volume	20 mL	[129; 647] mL	6, 5, 3, 2
End-systolic volume	20 mL	[63; 529] mL	6, 5, 3, 2
Mean aortic pressure	10 mmHg	[68; 121] mmHg	6, 5, 3
Peak-systolic aortic pressure	10 mmHg	[83; 182] mmHg	6, 5
End-diastolic aortic pressure	10 mmHg	[48; 99] mmHg	6, 5
Ejection time	50 ms	[115; 514] ms	6

Number of representative states As in the previous experiment, state space quantization was tuned based on eight scouting patients, which were later dismissed. The numbers of representative states (n_S) yielding the best scouting performance were 70, 150, 400 and 600 for the 2p, 3p, 5p and 6p setup, respectively.

Reference method A gradient-free optimizer [388] based on the simplex method was used as benchmark, where the sum of squared differences between computed and measured values, weighted by the inverse of the threshold in the convergence criteria to respect the different ranges of objective values ($\|\boldsymbol{c}\|_\psi$, cf. Eq. (5.3)), was minimized. To account for the different number of parameters $n_{\boldsymbol{x}}$, the maximum number of iterations was set to $50 \cdot n_{\boldsymbol{x}}$ for the different setups. With increasing number of parameters to be estimated, performance decreased (Fig. 5.10), as is expected due to the increasing problem complexity.

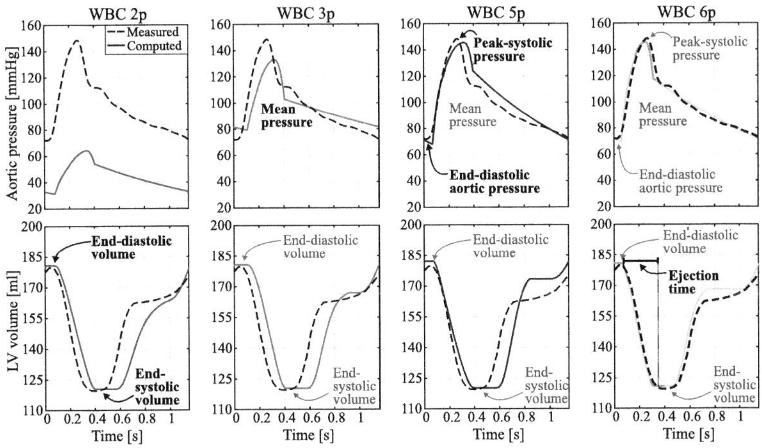

Figure 5.9. Goodness of fit (volume and pressure curves) after personalization of an example patient based on the different WBC setups. Additional objectives per setup are highlighted in bold. With increasing number of parameters and objectives, the proposed method manages to improve the fit between model and data.

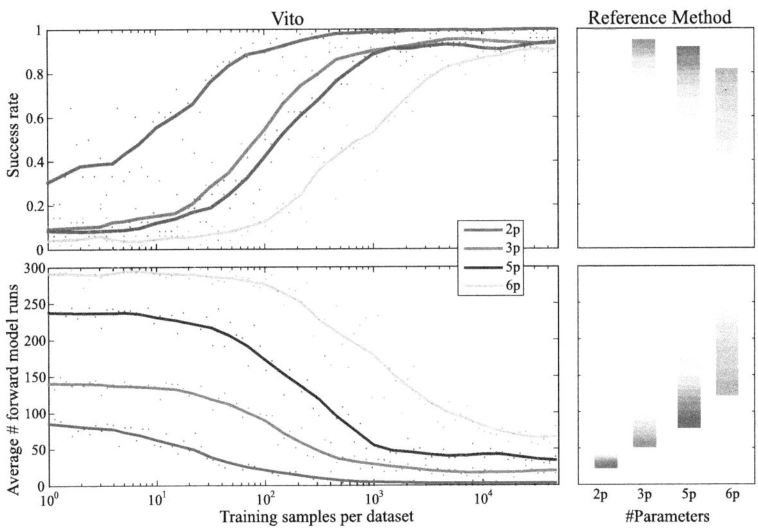

Figure 5.10. WBC personalization results (top: success rate, bottom: average number of forward model runs until convergence) for the different setups. Left: RL-based method performance over increasing number of training data (cross-validated personalization of all 48 datasets). Right: Performance of reference method. Each shade represents 10% of the results, sorted by performance; darkest shade: best 10%.

Convergence analysis For each WBC setup, the proposed method was evaluated based on varying amount of training data ($n_{\text{samples}} = 10^0 \ldots 10^5$) and using leave-one-patient-out cross-validation. We used the same dynamic limits for number of iterations as for the reference method. The results are presented in the graphs in Fig. 5.10. Performance increased steadily with the amount of data until reaching a plateau. The more complex the problem, i.e. the more parameters and objectives involved in the personalization, the more training data was needed to reach the plateau with best performance. As an example, 80% success rate was reached with less than $n_{\text{samples}} = 50$ training samples per dataset in the 2p setup, while for 6p, almost 90× as many samples were needed.

Compared to the benchmark, the RL-based method reached equivalent or better success rates (up to 11% higher), while simultaneously being more efficient in terms of run-time. In the 2p setup, if $n_{\text{samples}} \geq 10^3$, the RL-based converged after 3 iterations on average, as compared to 22.6 iterations for the reference method. This translates to a seven-fold speed-up. For all of the other setups (3p, 5p, and 6p), significant improvements can also be observed, with speed-up of at least a factor of 1.8.

5.4 Summary

This chapter presented two methods based on machine learning to estimate tissue properties. We first presented a method based on polynomial regression to estimate the electrical diffusivity of the myocardium from 12-lead ECG features, namely QRS duration and electrical axis. The approach is as accurate as traditional inverse problem methods, while it provides a first order approximation of parameter uncertainty and is extremely computationally efficient. The method, while illustrated on a mono-domain model of cardiac electrophysiology, is generic and could be applied to any type of cardiac electrophysiology models (mono-domain, bi-domain, Eikonal, etc.). Although the focus of this discussion was on cardiac depolarization and electrical diffusivity, a similar strategy could be used to estimate the tissue parameters related to cardiac repolarization, like action potential duration and restitution curve. Furthermore, the method can be enhanced by providing more features as input without loss of generality and hence reduce the uncertainty in estimated cardiac parameters.

One important learning from this approach is the importance of geometry in the data-driven approach. The relative position of the heart in the patient's torso has a noticeable impact on the ECG traces. It is therefore crucial to take the geometry out of the input features. In this work, we used evaluations of the forward model with specific values of the parameters to estimate the maximal range of model outputs for each patient, thus being able to normalize the ECG features and removing the expected effect of patient-specific geometry. Alternatively, geometric information could be used explicitly as input to the regression. More advanced neural networks models could also be employed to learn encodings of the geometric relationship between heart and torso, with the goal to further improve electrical diffusivity estimates.

We then described a personalization approach based on reinforcement learning. Inspired by how humans approach personalization, this method first attempts to understand the model characteristics by means of data-driven exploration. The obtained knowledge is then utilized to learn a personalization strategy using RL. We describe a generic solution that requires only minimal user input (parameter ranges, authorized actions, number of representative states) to start learning in an autonomous way how to personalize a model.

The approach was tested on two challenging personalization tasks in cardiac computational modeling. Only minimal tuning of the algorithm hyper-parameters was required, suggesting good generalization properties. At the same time, we observed increased robustness compared to state-of-the-art optimization approaches. Training data generation can be computationally demanding. However, training is performed only once in an off-line phase. On-line personalization, on the contrary, is not parallelizable in its current form, because, just like for most state-of-the-art methods, the parameters for each forward model run depend on the outcome of the previous iteration. However, we observed faster convergence with less evaluations of the forward model (which can be computationally demanding), even when compared to manually engineered algorithms. The on-line overhead introduced by the RL-based method (convert data into an MDP state, then query an action policy) is negligible.

Based on these promising results, we conclude that applying modern AI technology to model personalization could enable a unified framework for personalization of any computational physiology model. Eventually, such a framework may speed up progress in the computational modeling community as it could remove the need for tailored design and engineering of complex optimization techniques and promote the adoption of machine trained algorithms encoding the required domain-specific expertise through learning from data.

6

Additional clinical applications

Felix Meister[a,b], Helene Houle[c], Cosmin Nita[d], Andrei Puiu[d], Lucian Mihai Itu[d], Saikiran Rapaka[a]

[a]Siemens Healthineers, Princeton, NJ, United States. [b]Friedrich-Alexander University Erlangen-Nuremberg, Erlangen, Germany. [c]Siemens Healthineers, Ultrasound Division, Mountain View, CA, United States. [d]Siemens SRL, Image Fusion and Analytics, Brasov, Romania

6.1 Cardiac resynchronization therapy

6.1.1 Introduction

Heart failure (HF) is a cardiovascular disease (CVD) characterized by a reduced cardiac output due to a systolic or diastolic dysfunction [389,390]. Consequently, the heart cannot meet the body demand for oxygen and nutrients. Over the last decade, an increasing prevalence of HF has been observed, which has been linked to a growth of the elderly population and higher survival rates to acute myocardial infarction. To treat the failing heart, cardiac resynchronization therapy (CRT) proved to be efficient for patients that suffer from a dyssynchonous cardiac contraction caused by dilated cardiomyopathy (DCM). A set of electrodes is implanted to counteract arrhythmias and to resynchronize the beating cardiac chambers by emitting precisely timed electrical impulses. Several benefits associated with successful CRT treatment have been reported, including increased quality of life and reduced total mortality [391].

Even though CRT effectiveness has been demonstrated in several studies, it has also been observed that 30 to 50% of treated patients do not respond to this therapy [392–395]. Several factors have been identified, for instance of clinical nature or related to electrical properties of the diseased hearts. However, biomarkers that can reliably identify responders have not been identified so far [396–400]. Other related studies suggest that new imaging technologies like speckle tracking echocardiography [401–403] and cardiac MRI with late enhancement [404–406] may help in predicting the response prior to the intervention. In general, these approaches aim at improving the patient selection using baseline information, and hence neglect the effect of the procedure itself.

Artificial Intelligence for Computational Modeling of the Heart
https://doi.org/10.1016/B978-0-12-817594-1.00017-6

Importantly, it has been shown that the position of the electrode can significantly impact response to the therapy [391].

Personalized computational models have been used to complement clinical and observational studies to derive a deeper understanding of the disease and the effect of therapy [9]. For instance, Crozier et al. suggested that the location of the leads of the CRT device has a more significant impact than the presence of conduction blocks [407]. In [408], the authors concluded that the optimal left ventricular (LV) pacing site in left bundle branch block (LBBB) cases without scar should be most distant from the right ventricular (RV) lead. They further hypothesized that in cases with scar the LV lead needs to be placed remotely from both RV lead and scar. Willemen et al. validated in canine experiments the observations from computational models, indicating a different response of RV and LV function to changes in interventricular (VV) pacing delay [409]. In another study, Lee et al. used modeling to explore the long term effects of atrioventricular (AV) and VV delay optimization, suggesting decreased benefits over time [410]. Niederer et al. compared the benefits of multi vs single site LV pacing on virtual hearts, pointing out potential advantages in cases with posterolateral scar [411]. Finally, Costa et al. traced back the increased risk of VT in cases with scar to observations of increased repolarization dispersion around scar tissue [412], a situation that is becoming more and more noticeable clinically.

Such multi-scale models of the heart could ultimately represent a powerful tool for decision support, by enabling a physician to perform virtual CRT beforehand and assess the chances of response a patient has. Towards this aim, a cardiac modeling pipeline for CRT guidance based on pre-operative and non-invasive data is presented. Early results on a small cohort of patients are reported.

6.1.2 Methods

6.1.2.1 Data acquisition

The study comprised ten patients of age 18 and older that suffered from either dilated (90%) or ischemic (10%) cardiomyopathy and received CRT treatment following international recommendations at the Bordeaux University Hospital between January 2016 and October 2017 [413]. For each patient the following data acquisition protocol was applied.

At baseline, a 12-lead electrocardiogram (ECG) was recorded. Siemens Magnetom Aera with 1.5 Tesla (Erlangen, Germany) was used to acquire 2-chamber, 3-chamber, 4-chamber, and short axis cine images as well as delayed enhancement sequences, which were analyzed by radiologists to determine the scar burden per region of a 17 segment left-ventricular model as recommended by the American Heart Association (AHA). The affected AHA regions were hereby categorized based on scar characteristics as either intramural, epicardial, diffuse, or transmural scar and assessed at 25%, 50%, 75%, or 100%.

During the intervention, the AHA segment associated with the location of the left ventricular electrode and the septal location (apical, mid, or basal) of the RV electrode was reported [206]. Atrio-ventricular (AV) pacing delay and VV delay information was available additionally.

Three different configurations, namely LV only, RV only, or biventricular (BiV) stimulation, were tested in all patients, except one, resulting in a total of 28 stimulation protocols. The AV and VV delay per stimulation protocol was kept constant per patient. For validation, 12-lead ECGs were acquired.

6.1.2.2 Computational modeling

A personalized multi-physics computational model was estimated using the pipeline illustrated in Fig. 6.1. First the heart anatomy was segmented from the MR images as described in section 3.2 and refined in cases of under- and over-segmentation [31]. The endo- and epicardial segmentations given as surface triangulations were merged to form a closed surface and then tetrahedralized, resulting in a volumetric bi-ventricular myocardial model. A rule-based fiber model and mesh tagging was applied (refer to section 2.1). Furthermore, the scar burden per AHA segment, as per radiological report, was mapped to the model. Finally, a representative avatar of the human torso was fitted to the images, and ECG electrodes were placed consistently across patients according to clinical standards (section 2.1).

To compute the electrophysiological (EP) response, the graph-based model described in section 2.2.3 was applied. Left bundle branch block (LBBB) was modeled by disabling the associated left activation patway and constraining the left Purkinje system to have a conduction velocity equal or at most within 125% of the conduction velocity in the myocardium. The EP model was individualized following the approach presented in section 2.5.2.

For each patient the outcome from CRT treatment was predicted by simulating the stimulation protocols that had been previously tested on the patient. Left and right electrodes were placed as stated in the clinical report. Heart rate and pacing delays were also set accordingly. To cope with the uncertainty in the timing of the stimulations with respect to the spontaneous activation of the heart, which could potentially affect the presence of wave fusion,

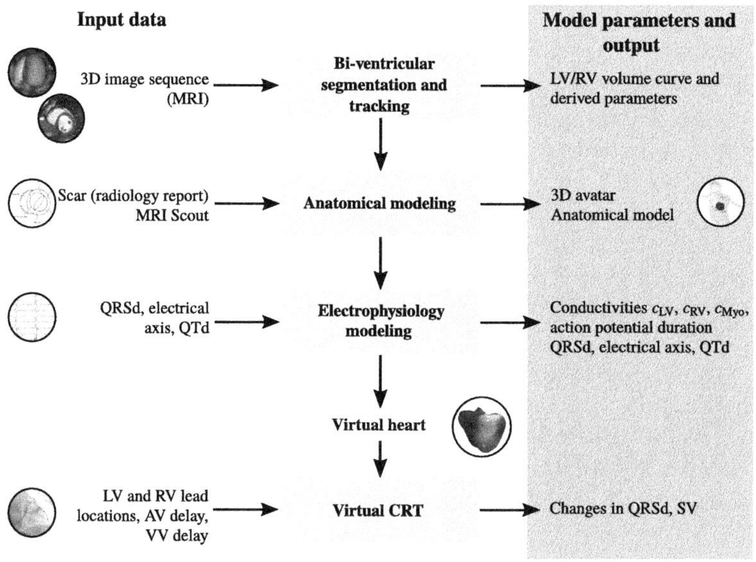

Figure 6.1. Illustration of the virtual CRT modeling pipeline from medical images and pre-operative, non-invasive measurements to the heart model.

the following procedure was applied. With the heart completely repolarized at the initial time t_{in} =0 ms, it was assumed that the depolarization of the ventricles starts at t_0 =160 ms (consistent with measurements of PR interval duration in humans [42]). The left ventricular electrode was set to pace at time t_{LV} = AV delay and the RV electrode at $t_{\mathrm{RV}} = t_{\mathrm{LV}}$+VV delay.

All experiments were performed on a standalone desktop machine with Intel Xeon E5 3.6 GHz CPU, NVidia Quadro M5000 GPU, and 32 GB of RAM. Using this configuration it took approximately 5 mins to automatically compute the anatomical model from the clinical images. The computation of electrical activation over one heart beat took 10 ms, which allowed to perform model individualization within 2 mins using traditional optimization methods. The objective of the optimization was minimization of the misfit between measured and computed ECG features (QRS interval duration and electrical axis, see also section 2.5.2).

6.1.3 Results

Model creation was successful in all cases. In detail, the measured electrophysiology at baseline conditions was matched with a mean absolute error in QRS duration of 2.8±8.4 ms and a median of 0 ms. The personalized model conductivities are reported in Table 6.1.

Table 6.1 Summary on the statistics of the personalized model parameters for EP.

Parameter	Mean ± SD (median)
c_{Myo} – mm/s	841 ± 137 (831)
c_{LV} – mm/s	1583 ± 1057 (1043)
c_{RV} – mm/s	3439 ± 825 (3999)

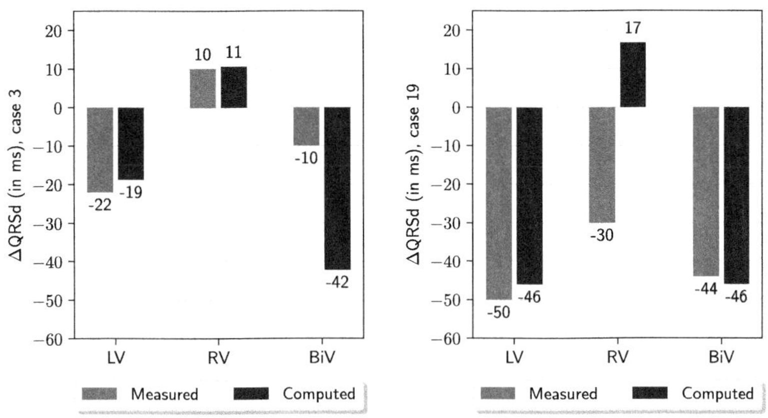

(A) (B)

Figure 6.2. Comparison of ΔQRSd measurements and predictions per stimulation protocol for (A) case 3 and (B) case 7.

6.1.3.1 Electrophysiological results

The observed variations in QRS duration varied per protocol in the studied population. Let $\Delta QRSd = QRSd_{Post} - QRSd_{Pre}$ denote the change in QRS duration observed under pacing with respect to the baseline (CRT OFF). In the population, QRS shortening could be observed for LV and BiV pacing (average $\Delta QRSd_{LV} = -5.3$ ms and average $\Delta QRSd_{BiV} = -8.8$ ms, respectively), whereas RV pacing saw an increase in QRS duration (average $\Delta QRSd_{RV} = 13.1$ ms). These observations were consistent with clinical knowledge. In terms of changes in QRS amplitude (i.e. absolute value of ΔQRSd), a change of 24±16 mms was observed throughout the protocols, with 20± 20 ms for RV pacing, 27±14 ms for LV pacing, and 24±20 ms for BiV pacing.

In 21 computational experiments out of the total 28 stimulation protocols the trend (defined as whether QRS shortened or not) was predicted correctly, resulting in an accuracy of 75%. Moreover, by defining QRS shortening post CRT as a positive

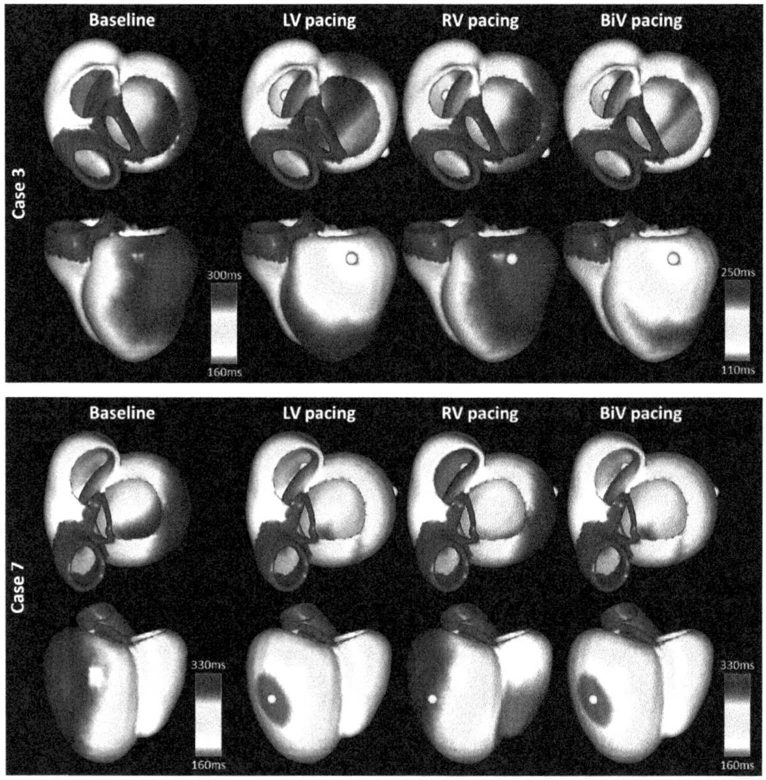

Figure 6.3. Illustration of electrical wave propagation for case 3 (upper) and case 7 (lower).

event, a ROC analysis could be performed for the accuracy of a test based on computed QRS shortening. The ROC curve was obtained by considering various thresholds τ for the computed ΔQRSd to define a shortening (ΔQRSd$< \tau$); the area under the curve (AUC) was 0.71, and the sensitivity and specificity for a zero threshold equaled 0.8 and 0.69, respectively. The model was able to predict QRS shortening under biventricular pacing in all 6 patients who saw an intra-procedural shortening of their QRS complex. Furthermore, the model was also able to detect QRS prolongation in two of the remaining four cases, while misclassifying the other two cases.

The capability of the model to predict the exact QRS duration was also evaluated. Over the entire population the absolute error in QRS duration post CRT treatment was 17$\pm$15 ms, with a Pearson correlation coefficient of 0.7. Similarly, the absolute error in ΔQRSd was 18$\pm$15 ms. In the case of biventricular pacing, an absolute error of 22$\pm$19 ms and 25$\pm$18 ms was observed for QRSd

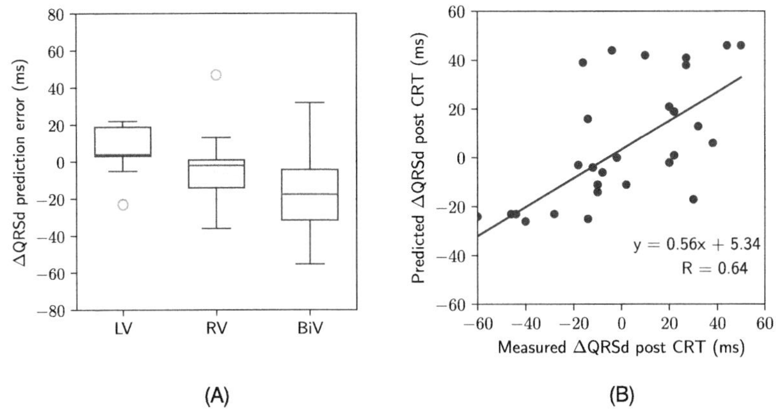

Figure 6.4. Quantitative analysis of predictive performance. (A) Mean ΔQRSd per stimulation protocol; (B) Measured vs. predicted ΔQRSd.

and ΔQRSd, respectively. In detail, we observed that the model performed better in predicting the outcome under LV and BiV pacing than in RV pacing (correlation coefficients of 0.9, 0.52, and 0.46, respectively). Furthermore, no significant bias in LV and RV prediction was apparent, but an over-estimation of QRS shortening was observed under biventricular pacing. A detailed comparison of measured and computed ΔQRSd for two representative cases can be seen in Fig. 6.2. Wave propagation for these two cases are illustrated in Fig. 6.3. (See Fig. 6.4.)

6.1.4 Discussion

This study aimed at a real-world scenario in which only baseline, standard of care data is available for computational modeling. Despite the limitations in the data, the changes in QRS complex duration for 75% of all protocols could be predicted. The model could also identify two out of four patients who did not see a shortening of their QRS complex under pacing. The computation of the exact electrophysiology post CRT was more challenging. This can be explained by limitations in the data, model and the lack of precise information about the CRT device. For instance, the delay between device stimulation and spontaneous stimulation had to be guessed, although it was found to significantly affect the QRS complex duration under pacing. Being able to run these models during the procedure, when more data is available, could prove valuable by enabling interactive model adjustment to the patient's physiology, for more precise predictions.

In summary, the feasibility of virtual CRT using patient-specific computational models was illustrated. The model could capture the trends in QRS shortening during pacing, and showed promising accuracy in terms of absolute QRS duration prediction. Extensions of this work could include evaluation on larger cohort, prediction of additional electrophysiology and hemodynamics biomarkers.

6.2 Aortic coarctation

6.2.1 Introduction

Coarctation of the aorta (CoA) is a congenital disease that accounts for 8–11% of cardiovascular diseases [414]. It consists of a narrowing of the aortic lumen, usually located near the aortic arch, that obstructs the flow, causing a significant pressure drop along the narrowing, and less blood being delivered to the lower part of the body. If left untreated, it can cause hypertension, early coronary artery disease, heart failure, stroke, aneurysm formation and decreased life expectancy [415–417]. Treatment of CoA aims at reducing the pressure drop and restoring normal flow conditions to the lower part of the body. Treatment is typically performed percutaneously by placing a stent or by balloon angioplasty, or surgically [418,419]. In older children and particularly in teenagers and adults, catheter-based intervention is becoming more prevalent and is typically the standard of care. Although the occlusion ratio can be accurately measured from image data, e.g. MRI or transthoracic echocardiography, accurate clinical decision making requires hemodynamic assessment that can be performed only invasively through pressure measurements [420]. Specifically, treatment is recommended in case of a resting state peak-to-peak trans-coarctation pressure drop of 20 mmHg or more [421]. The most accurate technique for measuring pressure drop is intravascular catherization.

Given the invasive nature of cardiac catheterization, and associated costs and risks, efforts have been made to develop non-invasive methods for computing the trans-CoA pressure drop from medical images. A previously reported approach consists in using flow velocity information extracted with Doppler ultrasound or phase contrast MRI (PC-MRI), in combination with fluid dynamics equations, to estimate pressure as a function of velocity [422–424]. This approach tends to overestimate the pressure drop, and is not suitable for older children or adults due to the posterior location of the descending aorta. Although PC-MRI is able to provide significantly higher quality velocity measurements [425,426], the imaging time is typically 10–15 minutes [427], too high for a routine clinical environment. A simple non-invasive alternative consists in performing blood pressure measurements in the upper and lower body extremities, and computing the difference [428].

As described in section 2.4, a different approach relies on Computational Fluid Dynamics (CFD) for performing patient-specific blood flow computations. Compared to traditional techniques, the main advantage of a CFD based tool is that it offers the potential to also simulate post-treatment hemodynamics or flow under different conditions such as stress, which would normally have to be induced pharmacologically. Furthermore CFD is able to provide quantities that are not available through measurement, e.g. Wall Shear Stress (WSS) or WSS-derived quantities. As computing power and technology improved over the years, CFD has been used increasingly as a research tool for studying aortic flow under patient-specific conditions and different physiological assumptions: rigid vessel walls [429,430] or compliant walls [431–433], personalized inlet boundary conditions [434], idealized [433] or reduced order CFD modeling [435,436], etc.

CFD based approaches were also proposed for assessing aortic coarctation patients. These methods typically rely on medical images acquired through either magnetic resonance imaging (MRI) [437] or Computed Tomography (CT) [229], in addition to some physiological parameters, such as cuff blood pressures or echocardiographic flow velocities.

Itu et al. [436] proposed a non invasive workflow, based on MRI data, for hemodynamic assessment of aortic coarctation, relying on a reduced-order CFD model coupled with a pressure drop model for the CoA segment. The pressure drop model was based on the work previously published by Young and Tsai [438], and integrated into the reduced-order model for accurate computation of energy losses at the CoA site. However, the pressure drop model was designed and developed based on idealized stenotic shapes, which do not reflect the complexities and anatomical variabilities of aortic coarctation.

As machine learning (ML) evolved and was successfully applied to solving more complex problems, it also gained popularity as a technique for accelerating/replacing complex computations that would normally require long execution times such as computational fluid or structural dynamics. Liang et al. [439] proposed a machine learning approach for replacing computationally demanding Finite Element Method (FEM) computations for predicting surface stress distribution in aortic vessels. Specifically, they performed a large number of FEM computations on synthetically generated aortic vessels, and then trained a neural network to directly predict the surface stress distribution using only the vessel shape as input. Similarly, Liang et al. [440] proposed a machine learning approach for investigating the relationship between shape features and risk of rupture for aortic aneurysm. This methodology has also been employed for accelerating CFD computations: Hennigh [441] introduced a Deep Neural Network architecture for reducing time and memory usage in Lattice Boltzmann flow computations, and Tompson et al. [442] used Convolutional Neural Networks for accelerating fluid computations on Eulerian grids.

In this section, we introduce a ML based pressure drop model capable of accurately determining energy losses for a wide range of flow conditions and anatomical CoA variations. Inspired by the original work of Young and Tsai [438], where model parameters were fitted to experimental data, the proposed ML based pressure drop model was developed using a similar approach, but relying on in silico data. A large number of 3D CFD computations were performed on a set of synthetically generated aortic coarctation anatomical models, and the resulting data was used to train a ML based model for accurate CoA pressure drop prediction. A similar approach was described in section 4.1 for estimating FFR in coronary arteries. However in this case three-dimensional CFD simulations were used instead of 1D, therefore fundamentally changing the process of generating synthetic data and requiring a lot more computationally intensive CFD simulations.

6.2.2 Methods

6.2.2.1 Generation of a synthetic training database

The process of generating synthetic CoA anatomical models consists in fitting a parameterized surface model to a set of patient-specific models and using the fitted model to generate synthetic CoA geometries.

In a first step, the given patient-specific anatomical models are processed so that only the region of interest (descending aorta containing the CoA segment) is retained and the resulting meshes have the same orientation, scaling and position. The meshes are then rotated such that the x axis becomes the direction of maximum extent and the y axis is the secondary direction of maximum extent, as identified through principal component analysis on the mesh points position. While mesh cutting is performed manually, the rotations are performed automatically by using the computed principal component vectors. Finally, the meshes are normalized to ensure they are centered in the origin and cover a range between -1 and 1. The main purpose of rotation and normalization is to maximize the overlap between different samples, since rotation angles and absolute position in space are irrelevant to the results of CFD computations. Fig. 6.5 displays the workflow for mesh pre-processing.

The synthetic vessel model consists of two separately parameterized functions: one for describing the vessel centerline, and one for describing the surface. The vessel centerline is a three-dimensional Bezier curve defined by five control points, while the surface model is a function $\bar{r} = \bar{r}(z, \phi)$, where r is the vessel radius, z is the position on the centerline, and ϕ is the angle. The surface points are expressed in cylindrical coordinates with respect to the vessel centerline, as displayed in Fig. 6.6

The Cartesian (x, y, z) to cylindrical coordinates (z, r, ϕ) transformation allowed us to express the radius, the most relevant characteristic for the CoA segment, as an explicit function of two variables (z and ϕ), i.e. represent it as a three-dimensional explicit surface. The main goal of this step is to fit a function $\bar{r} = \bar{r}(z, \phi)$ to the given points, to obtain a parameterized encoding of the surface. The transformation from Cartesian to cylindrical coordinates is performed in two steps:

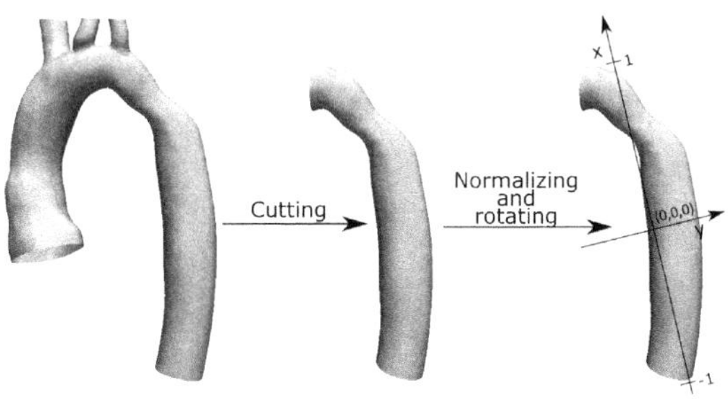

Figure 6.5. Workflow for pre-processing the patient-specific anatomical models.

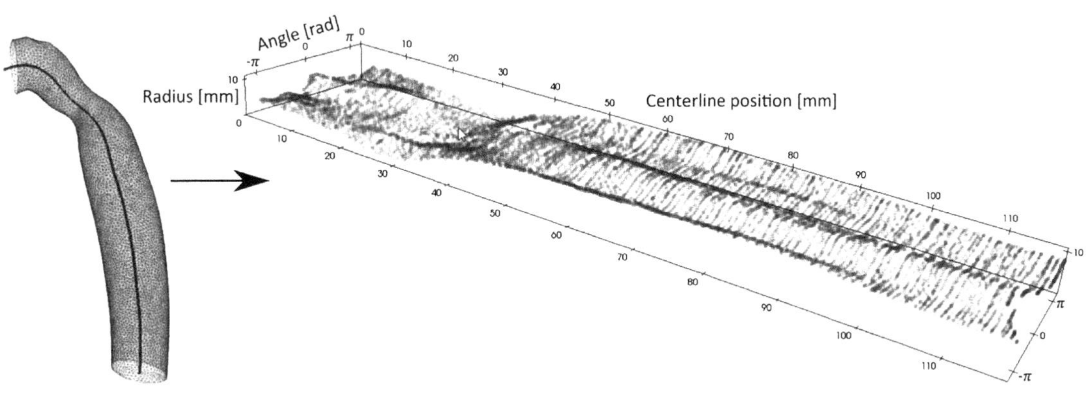

Figure 6.6. Transforming 3D surface points to cylindrical coordinates, with respect to the centerline position.

1. The points are expressed in Frenet coordinates with respect to the centerline. For each point $\mathbf{p}$ on the surface, its corresponding projection on the centerline $\mathbf{p}_\perp$ is found, and the centerline normal $\mathbf{n}$ and bi-normal $\mathbf{b}$ vectors are computed at point $\mathbf{p}_\perp$. The surface point is transformed to the local coordinate system as follows: $\mathbf{p}_F = (z, \mathbf{p} \cdot \mathbf{n}, \mathbf{p} \cdot \mathbf{b})$ where z is the arc length of the centerline segment given by the centerline starting point and the projected point $\mathbf{p}_\perp$.
2. The radius r and the angle ϕ are computed by transforming point $\mathbf{p}_F$ to cylindrical coordinates:

$$r = \sqrt{(\mathbf{p} \cdot \mathbf{n})^2 + (\mathbf{p} \cdot \mathbf{b})^2}, \tag{6.1}$$

$$\phi = arcos\left(\frac{\mathbf{p} \cdot \mathbf{n}}{\mathbf{p} \cdot \mathbf{b}}\right). \tag{6.2}$$

The angle is then shifted to the interval $(-\pi, \pi)$, depending on the quadrant of $\mathbf{p}_F$, i.e. $\phi := -\phi$ if $\mathbf{p}_F$ is in the 3rd or 4th quadrant.

Once the points are expressed in (z, r, ϕ) coordinates, a surface fitting operation is performed. The model function is chosen such that (i) it is able to capture only the relevant geometric features of the vessel, and (ii) each feature is directly controllable by one parameter.

$$\bar{r}(z, \phi) = B(z)S(z, \theta_{surface}), z \in [0, 1], \tag{6.3}$$

$$B(z) = \sum_{i=1}^{5} b_{i,5}(z)r_i, \tag{6.4}$$

$$S(z, \phi) = 1 - \left(\left(1 - s_1 cos^2\left(\frac{\phi}{2}\right)\right) s_2 e^{-s_3(z-s_4)^2} - s_5 e^{s_6(z-s_7)^2}\right), \tag{6.5}$$

where $b_{i,5}(z)$ are Bernstein basis polynomials of degree 5 and $\theta_{surface} = [r_1, r_2, \ldots, r_5, s_1, s_2, \ldots, s_5]$ are the parameters to be optimized. The surface model $\bar{r}$ is composed of two separate components B and S (see Fig. 6.7):

1. B is a one dimensional Bezier curve, defining the main vessel tube. It captures features corresponding to the entire anatomy, e.g. bulk radius, inlet to outlet tapering, and other small variations.
2. S is composed of two Gaussian functions, defining the region of the CoA narrowing. The first exponential term corresponds to the actual narrowing part of the vessel, while the second exponential term corresponds to the "bulging" part, which is

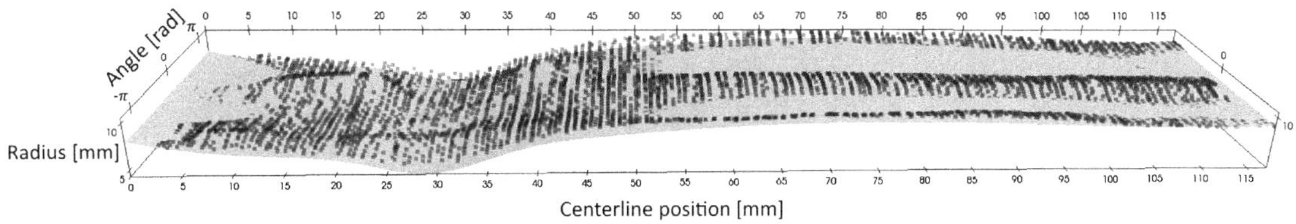

Figure 6.7. Fitting the surface model $\bar{r}(z,\phi)$ to the points corresponding to the patient-specific anatomical model.

typically present immediately downstream from the CoA. The $(1 - s_1 cos^2(\phi/2))$ factor of the first exponential term models the CoA eccentricity, i.e. if the factor s_1 is non zero the CoA is no longer axi-symmetric.

The procedure described above is applied only for defining the radius as a function of centerline position and angle, i.e. it provides no information about the centerline of the vessel (curvature, torsion, etc). An additional model function is employed for the vessel centerline, consisting of a three-dimensional Bezier curve containing 5 control points:

$$\mathbf{C}(t) = \sum_{i=1}^{5} b_{i,5}(z)\mathbf{c}_i,\ z \in [0, 1], \tag{6.6}$$

where $\theta_{cl} = [\mathbf{c}_1, \ldots, \mathbf{c}_5]$ are the control points, representing the parameters to be optimized.

The model fitting process is formulated as a least squares problem, and the cost function is minimized using the optimize package in the Scipy Python library [443]. The fitting process is run separately for the surface and the centerline model, using the patient-specific anatomical models, so that a set of $\theta_{surface}$ and θ_{cl} parameters are obtained for each given patient-specific model. To fit the centerline model, the given points should be normalized, centered in the origin, and have the same orientation. Otherwise the parameter values vary significantly from case to case, therefore making it impossible to generate realistic synthetic centerlines. Besides the parameters described above, an additional parameter θ_{scale} is required for de-normalization, i.e. to scale the synthetic anatomical models to physiological ranges. The θ_{scale} parameters are computed for each given patient-specific model in the pre-processing stage, when the meshes are scaled. Thus, the final parameter list θ consists of the surface parameters $\theta_{surface}$, the centerline parameters θ_{cl} and the scaling factor θ_{scale}.

The synthetic models are generated using Eq. (6.3), and (6.6) by randomly choosing the θ parameters in the value ranges deter-

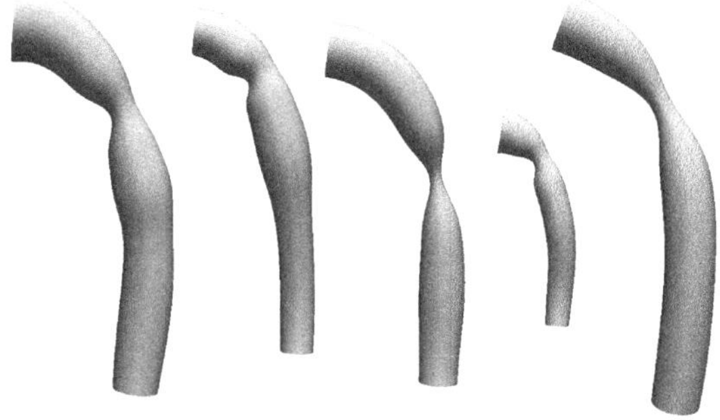

Figure 6.8. Examples of synthetically generated CoA anatomical models.

mined for the patient-specific cases. Eq. (6.6) is employed to generate centerline points in Cartesian coordinates, while Eq. (6.3) is used to generate surface points in cylindrical (r, z, ϕ) coordinates. Unreported experiments suggest that best results are achieved when parameter sampling is conducted following a normal distribution. Since the model is formulated as a set of continuous functions, the surface can be generated with an arbitrary number of points. For generating a synthetic anatomy, the surface function is evaluated at a finite number of uniformly distributed points based on which a polygonal mesh is then defined. To increase the anatomical fidelity of the resulting surface, some additional constraints are defined for:

1. The ratio between inlet and outlet radius (0.8–2).
2. The ratio between the healthy radius (upstream from the CoA segment) and the CoA minimum radius (0.2–0.8).

Finally, 100 points along the centerline, and 50 points along each cross section are generated and transformed from cylindrical to Cartesian coordinates, obtaining thus the actual surface points used to generate the final surface mesh. (See Fig. 6.8.)

6.2.2.2 Three-dimensional flow computations

The 3D flow model is based on the Lattice Boltzmann Method (LBM), which describes physics of fluid flow at a mesoscopic scale by taking into account molecular interactions between flow particles. The LBM provides ultimately the same solution as the Navier–Stokes based solvers [214], but it is inherently highly parallelizable, which is a necessary feature to enable fast computation for the large database of synthetic geometries.

Aortic flow typically has a high Reynolds number (up to 5000 at peak systole); to simulate this flow regime with LBM without producing numerical instabilities, the spatial resolution needs to be very small, resulting in very high computational cost. Alternative formulations of the method, such as the entropic Lattice-Boltzmann method (ELBM) [444] can be considered to reduce the computational complexity of simulating physiological aortic flow.

Since LBM operates on a Cartesian grid and the 3D model of the aorta is a surface mesh, it is embedded in the grid by computing a signed distance function $\phi(\mathbf{x})$ for each location $\mathbf{x}$ in the grid, i.e. $\phi(\mathbf{x})$ is the signed distance from the grid node $\mathbf{x}$ to the closest point on the surface mesh. The sign is computed such that $\phi(\mathbf{x}) \leq 0$ for grid nodes located inside the vessel, and $\phi(\mathbf{x}) > 0$ for nodes located outside of the vessel, hence the surface location can be accurately identified as $\phi(\mathbf{x}) = 0$. The surface mesh also contains a label associated to each polygon, depending on the boundary condition that needs to be applied, i.e. inlet, outlet or no-slip wall. The surface labels are associated to each grid node along with ϕ during the pre-processing step. Thus, the boundary nodes can be identified, and the type of boundary condition that needs to be applied is determined.

Since a large number of flow computations are required, performance becomes an important aspect. Being LBM readily parallelizable, it is possible to resort to optimized GPU-based implementations [445–448].

The purpose of the flow computations is to find a correlation between geometric features of the vessel and pressure drop. All computations were performed under a steady state configuration, i.e. a constant flow rate was imposed at the inlet, and the flow computation was run until convergence was reached. Unsteady flow computations were shown to have a limited impact on global hemodynamic quantities, including pressure drop, at the expense of very time-consuming computations [429,449].

For each synthetic surface a total of six computations were performed, varying the inlet Reynolds number between 500 and 5000. To ensure that the resulting flow rate and pressures lie in a physiologically plausible range, the inlet flow rate was constrained to be smaller than 32 L/min, and the resulting pressure drop below 60 mmHg. Since the pressure drop cannot be determined a priori, it was estimated using the Young-Tsai model [438] as a function of inlet flow rate and the CoA degree. Thus, the inlet flow rates for a case were chosen as follows:

1. The initial inlet flow rates are computed such that the Reynolds numbers are 500, 1000, 2000, ..., 5000 respectively:

$$Q = \frac{Re A_{inlet} \nu}{2 r_{inlet}}, \tag{6.7}$$

where A_{inlet} and r_{inlet} are the inlet area and radius respectively, and ν is the kinematic viscosity of blood.

2. For each of the computed flow rates, the pressure drop ΔP is estimated using the Young-Tsai model. If the maximum ΔP (corresponding to the largest Reynolds number) is larger than the physiological threshold of 60 mmHg, all the Reynolds numbers are scaled down until the maximum ΔP becomes 60 mmHg.
3. If any of the computed flow rates is larger than 32 L/min, the computation corresponding to that flow rate is not performed.

6.2.2.3 Pressure drop model for aortic coarctation

The pressure drop model initially proposed by Young and Tsai [438] is formulated as a function of flow rate containing a linear and a quadratic term that account for the viscous and turbulent effects respectively. In its initial form, the Young-Tsai model is formulated as follows:

$$\Delta P_{YT} = \frac{\mu K_v}{2\pi r_0^3} Q + \frac{\rho p_4}{2A_0^2}\left(\frac{A_0}{A_s} - 1\right)^2 Q^2, \tag{6.8}$$

$$K_v = p_1 \frac{L_a}{r_0}\left(\frac{A_0}{A_s}\right)^2, \tag{6.9}$$

$$L_a = p_2 L_s + p_3 r_s, \tag{6.10}$$

where, μ is the dynamic viscosity of blood, r_0 and r_s are the healthy and stenosed radius respectively, A_0 and A_s are the healthy and stenosed areas, L_s is the stenosis length and Q is the flow rate. Finally, the factors $p_1 \dots p_4$ are regression parameters, and were determined by fitting the model to experimental data. Young and Tsai performed the experiments on idealized, in vitro, stenosed vessel geometries by applying different flow rates. The numerical values for these parameters are given in Table 6.3.

Our work aims at improving the original Young-Tsai model by specializing it for aortic flow conditions. Thus, we explored three strategies: (i) a fine-tuning of the regression parameters, (ii) adding additional terms to Eq. (6.8) and finally (iii) coupling the resulting optimized model with a deep neural network.

The resulting pressure drop model was formulated based on Eq. (6.8) and three additional terms as follows:

$$\Delta P = \Delta P_{YT} + p_5 \Delta P_{convective} + p_6 E + p_7 B, \tag{6.11}$$

where, p_5, p_6 and p_7 are the parameters that determine the contribution of the additional terms to the total pressure drop.

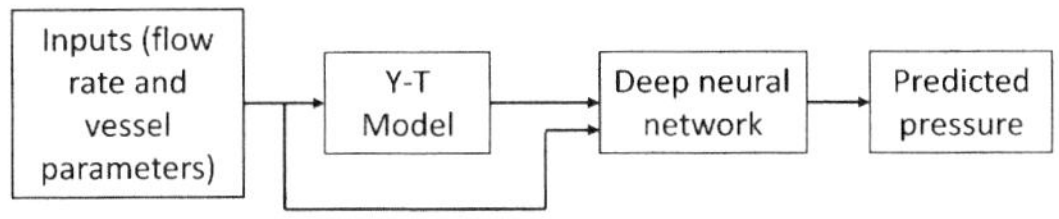

Figure 6.9. Cascaded pressure drop model resulting from coupling the optimized Young-Tsai model with a deep neural network. The neural network is used as a correction to the Young-Tsai model and it predicts the pressure based on both the Y-T model output but also the input quantities.

The second term of Eq. (6.11) accounts for the convective pressure drop given by the radius change across the CoA segment. Specifically, the convective pressure drop term is based on the work of Huo et al. [450], and it accounts for the energy loss given by the constriction of the cross-sectional area:

$$\Delta P_{convective} = \frac{\rho Q^2}{2}\left(\frac{1}{A_{outlet}^2} - \frac{1}{A_{inlet}^2}\right), \tag{6.12}$$

where A_{inlet} and A_{outlet} are the inlet and outlet areas respectively.

The last two terms of Eq. (6.11) account for geometric features that were found to be correlated with the total pressure drop. The term p_6E accounts for the pressure drop given by CoA eccentricity (parameter E corresponds to ϕ in Eq. (6.5)). The term p_7B represents the "bulging" part that is present downstream from the CoA, and parameter B corresponds to parameter s_5 in Eq. (6.5).

As for the coupled deep neural network component, it consists of 7 dense layers using relu activation functions and containing a total of 3210 parameters. The network is coupled with the Y-T model outputs but also directly with the inputs (flow-rate, vessel geometry parameters, etc). Fig. 6.9 displays the resulting model.

For evaluating the pressure model (Eqs. (6.8) to (6.12)) on a patient specific case, the values of the following geometric parameters need to be determined: K_v, L_a, L_s, E and B. For their computation, the surface model given by Eq. (6.5) is fitted to the patient-specific aortic surface.

A joint optimization of the parameters p_i and the neural network parameters can be performed so that the final, coupled model, best fits the 3D CFD pressures. In our experiments, the input features in the training database were flow rate, inlet area, minimum area (located at the CoA), outlet area, CoA length, eccentricity, bulging, and the ground truth quantity was the computed pressure drop. To capture the non-linear effects present at high Reynolds number flow, e.g. pressure recovery downstream from the CoA, the pressure drop was extracted as the total (inlet–outlet) pressure drop.

Table 6.2 Quantitative comparison of the pressure drop models: Norms of differences between 3D CFD and analytically estimated pressure drop.

	L_1	L_2	L_∞	Pearson
Original Y-T model	6.95	142.09	66.66	0.891
Optimized model	2.53	17.1	16.38	0.981
Coupled model	1.28	3.66	5.67	0.984

Training was performed on a subset containing 80% of the samples while the remaining 20% were used for testing. Splitting was performed at case level, to ensure that all computations performed for the same synthetic case, always lie either in the training or in the test set. Training was performed using the Tensorflow library [451] and the Adam optimizer [452], using mean absolute error as loss function.

6.2.3 Results

6.2.3.1 Evaluation of the pressure drop model

First we evaluate the performance of the optimized and the original Young-Tsai model using the 3D CFD computations as ground truth: Fig. 6.10 displays the results on the test set. Furthermore, Table 6.2 displays the results of the quantitative comparison of the three models. The original model underestimates the pressure drop since the synthetic CoA models have a higher degree of complexity than the original geometries used for fitting Eq. (6.8), and the range of considered Reynolds numbers is also larger. The herein introduced DL-based pressure drop model reduces the mean absolute error from 6.95 to 1.28 mmHg when comparing results against 3D CFD. Parameters p_1, p_2 and p_3 from the optimized model are related to the viscous pressure loss: p_1 decreases significantly in the optimized model, but the reduction is partially compensated by the increase of p_2 and p_3. Parameter p_4, which is related to the turbulent pressure loss, is modified only marginally, while parameters p_5, p_6 and p_7 are all non-zero, indicating that convection, eccentricity and bulging are relevant features for the overall trans-coarctation pressure drop. The errors are further reduced by the coupled model, therefore providing the best prediction. This suggests that the pressure drop is significantly affected by highly non-linear effects which are only captured by the neural network model.

Table 6.3 Pressure drop model parameters as they appear in Eqs. (6.8)–(6.11): p_1, p_2 **and** p_3 **are related to the viscous pressure loss,** p_4, p_5, p_6 **and** p_7 **are related to the effects of turbulence, convection, eccentricity and respectively bulging on the pressure loss.**

	p_1	p_2	p_3	p_4	p_5	p_6	p_7
Original Y-T model	0.83	3.28	16.0	0.9	–	–	–
Optimized model	0.114	8.776	19.743	0.980	1.485	0.327	0.119

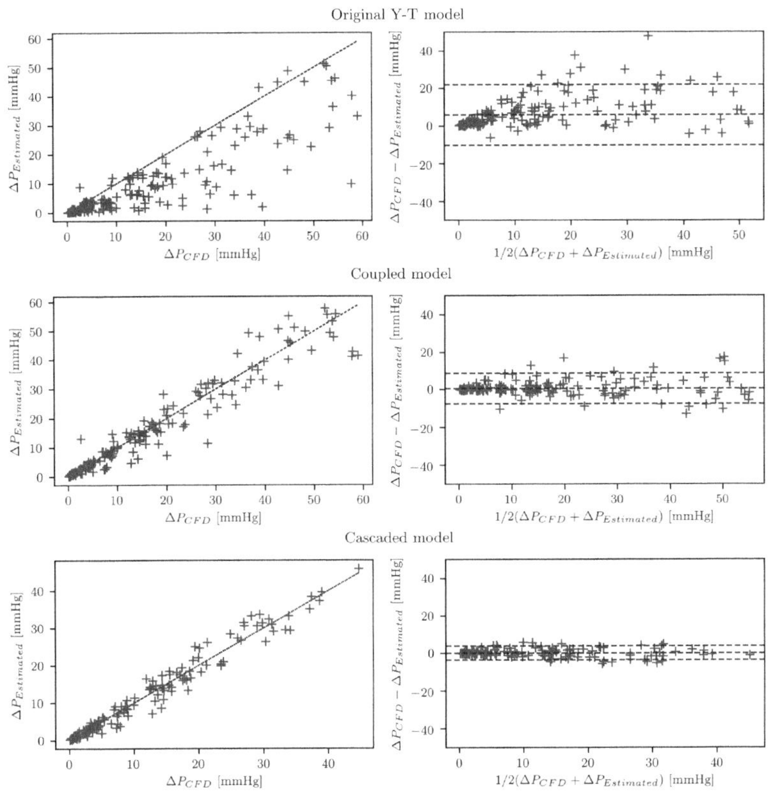

Figure 6.10. Evaluation of the pressure drop models. ΔP_{CFD} represents the pressure drop extracted from the 3D CFD computations, while $\Delta P_{Estimated}$ is the pressure drop determined analytically using the pressure drop models: top – original Young-Tsai model (Eq. (6.8)), middle – optimized pressure drop model and bottom – coupled model. The line in the scatter plot is the $y = x$ line.

6.2.4 Discussion

We have presented a machine learning approach for predicting the pressure drop of aortic coarctation. A comprehensive training database is an important prerequisite for the successful development of such a model. In an ideal scenario, the training database would consist of thousands of anatomical models extracted from medical images, accounting for the variability of aortic coarctations across different patient populations, and the corresponding invasive pressure measurements of each narrowing. From a practical point of view, establishing such a large database would be prohibitively expensive and time-consuming.

To address this issue, we reused the concepts introduced in [334]: a training database was generated consisting of synthetic aortic geometries, and corresponding pressure drop values computed from a CFD algorithm. The synthetic database is parameterized by the morphological features of the aortic configuration, allowing for a proper sampling of a relatively large range of anatomical configurations, addressing eccentricity, bulging, curvature, etc. After training, the machine learning algorithm encodes the correlation between the set of chosen geometric features and the quantity of interest, herein pressure drop, predicted by the validated 3D CFD model. The considered geometric features are highly specialized for pressure loss prediction, being related to viscosity and turbulence (as present in the original Young-Tsai model), convection, eccentricity and bulging. As can be observed in Table 6.3, after model optimization all seven pressure loss coefficients are non-zero, indicating that the CoA pressure drop is significantly influenced by each of the considered features.

The results here presented show the superiority of the proposed deep-learning based pressure drop model compared with a traditional one, in approximating pressure drop values produced by a 3D CFD model in a database of synthetically generated vascular geometries representative of CoA anatomies. The performance of the learning based pressure drop model could be further improved by increasing the number of synthetic anatomical models and the corresponding number of computations performed with the 3D CFD model. The set of features could also be further expanded to include other characteristics of the coarctation, e.g. aorta curvature in the CoA region, non-circular vessel cross-section, etc.

It should be noted that the presented approach is generic with respect to the CFD model used for generating the ground truth information. For this work, we used a 3D CFD model based on the Lattice Boltzmann Method. On the other hand, the accuracy of the pressure drop model compared to invasively measured pressures will depend on the accuracy of the CFD model used in the training phase, which was not addressed in this discussion. Since the proposed machine learning based pressure drop model learns the output of a computational blood flow model, it inherits the limitations of the blood flow model. The definition of the parameters of the blood flow models (both reduced-order and 3D CFD) relies on physiological assumptions, which would also require validation on larger data sets. Such assumptions include for example the mechanical behavior of the vascular wall (rigid, elastic, viscoelastic), the use of population-averaged rheological properties of the blood, and so on.

6.3 Whole-body circulation

6.3.1 Introduction

Due to the prohibitive computational cost of spatially detailed blood flow models (three-dimensional models in particular), closed loop models of the cardiovascular system rely heavily on lumped parameter modeling techniques, which are based on the analogy between hydraulic and electrical systems. The whole body circulation (WBC) model employed herein, displayed in Fig. 6.11, contains a heart model (left ventricle (LV) and atrium, right ventricle and atrium, valves), the systemic circulation (arteries, capillaries, veins), and the pulmonary circulation (arteries, capillaries, veins) [453]. Chapter 3 has described established techniques for cardiac image parsing and structure tracking. The anatomical information extracted using these methods, e.g. time-varying ventricular volume, may then be used as input for blood flow computations, as described more extensively in section 2.4.

While the lumped parameter model is computationally very efficient (a single forward run requires a runtime of milliseconds), its personalization requires hundreds of forward runs, leading to an overall computation time of 30 to 60 seconds for determining the patient-specific measures of interest. To remove the need for that computationally intensive step, we propose to estimate those measures using a deep neural network. This approach leads to a significant acceleration of computation time, thus making this modeling framework suitable for use in clinical workflows even within an intra-operative setting, where decisions need to be taken in near real-time.

6.3.2 Methods

6.3.2.1 Traditional personalization framework

The WBC model may be run under patient-specific conditions to compute various clinically relevant measures of interest: arterial resistance, arterial compliance, dead volume of the left/right

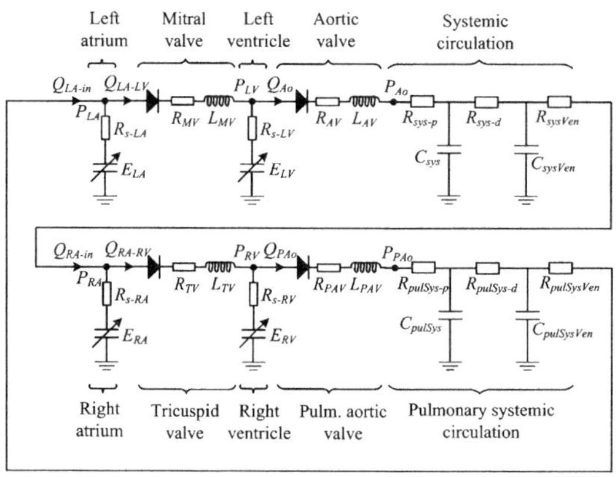

Figure 6.11. Lumped parameter closed loop model of the cardiovascular system. *Notations:* Q_{LA-in}: left atrial inflow; P_{LA}: left atrial pressure; Q_{LA-LV}: flow rate through the mitral valve; P_{LV}: left ventricular pressure; Q_{Ao}: aortic flow rate; P_{Ao}: aortic pressure; Q_{RA-in}: right atrial inflow; P_{RA}: right atrial pressure; Q_{RA-RV}: flow rate through the tricuspid valve; P_{RV}: right ventricular pressure; Q_{PAo}: pulmonary artery flow rate; P_{PAo}: pulmonary artery pressure; R_{s-LA}: left atrial source resistance; E_{LA}: left atrial time-varying elastance; R_{MV}: mitral valve resistance; L_{MV}: mitral valve inertance; R_{s-LV}: left ventricular source resistance; E_{LV}: left ventricular time-varying elastance; R_{AV}: aortic valve resistance; L_{AV}: aortic valve inertance; R_{sys-p}: proximal systemic resistance; C_{sys}: systemic compliance; R_{sys-d}: distal systemic resistance; R_{sysVen}: venous resistance; C_{sysVen}: venous compliance; $Rs - LA$: right atrial source resistance; E_{RA}: right atrial time-varying elastance; R_{TV}: tricuspid valve resistance; L_{TV}: tricuspid valve inertance; R_{s-RV}: right ventricular source resistance; E_{RV}: right ventricular time-varying elastance; R_{PAV}: pulmonary valve resistance; L_{PAV}: pulmonary valve inertance; $R_{pulSys-p}$: proximal pulmonary resistance; C_{pulSys}: pulmonary compliance; $R_{pulSys-d}$: distal pulmonary resistance; $R_{pulSysVen}$: pulmonary venous resistance; $C_{pulSysVen}$: pulmonary venous compliance.

ventricle, stroke work, ventricular/atrial/arterial elastance, arterial ventricular coupling, pressure-volume loop, etc. Thus, model parameters need to be personalized to match the patient-specific conditions and state [454].

The personalization framework used herein has been previously described in detail [151], and consists of two sequential steps. First, a series of parameters are computed directly (average aortic flow rate, total systemic resistance, total pulmonary resistance), and next, a fully automatic optimization-based calibration method is employed to estimate the values of the remaining parameters (see time-independent quantities below), ensuring that the personalized computations match the measurements.

The patient-specific input parameters are:

- **Systemic circulation:** peak aortic systolic pressure, end-diastolic aortic pressure, left ventricular end-systolic and end-diastolic volumes, left ventricular ejection time.
- **Pulmonary circulation:** peak pulmonary artery systolic pressure, end-diastolic pulmonary artery pressure, right ventricular end-systolic and end-diastolic volumes, right ventricular ejection time.

The personalized measures of interest determined after running the personalization are:

- Time-independent quantities:
  - **Systemic circulation:** dead volume of the left ventricle, time at maximum left ventricular elastance, systemic resistance, systemic compliance, ratio of proximal to distal resistance of systemic circulation.
  - **Pulmonary circulation:** dead volume of the right ventricle, time at maximum right ventricular elastance, pulmonary resistance, pulmonary compliance, ratio of proximal to distal resistance of pulmonary circulation.
- Time-dependent quantities:
  - **Systemic circulation:** aortic, left ventricular and left atrial pressures, left ventricular volume, aortic flow rate, and left ventricular pressure-volume loop.
  - **Pulmonary circulation:** pulmonary artery, right ventricular and right atrial pressures, right ventricular volume, pulmonary artery flow rate, and right ventricular pressure-volume loop.

The fully automatic optimization-based calibration method is formulated as a numerical optimization problem, the goal of which is to find a set of parameter values (initial volume in the close loop, left ventricular dead volume, time at maximum left ventricular elastance, systemic resistance, systemic compliance, ratio of proximal to distal resistance of systemic circulation, dead volume of the right ventricle, time at maximum right ventricular elastance, pulmonary resistance, pulmonary compliance, ratio of proximal to distal resistance of pulmonary circulation) for which a set of objectives is met (left ventricular average, minimum, and maximum pressure, left ventricular minimum and maximum volume, aortic valve ejection time, right ventricular average, minimum, and maximum pressure, right ventricular minimum and maximum volume, pulmonary valve ejection time). The number of parameters to be estimated is set equal to the number of objectives, and, thus, the parameter estimation problem becomes a problem of finding the root for a system of nonlinear equations. To solve the system of equations, we use the dogleg trust region method [151].

6.3.2.2 Deep learning based personalization

Setting up a large medical dataset is often infeasible due to the costs. As a first step, a database of 10 000 synthetically generated data samples has been created, reflecting the anatomical and functional variations representative of healthy and pathological cases. For the systemic circulation, the following input parameters were randomly chosen: heart rate, systolic blood pressure, diastolic blood pressure, left ventricular end-diastolic volume, stroke volume, and left ventricular ejection time. The cycle-averaged blood pressure was computed from the systolic and diastolic blood pressures and from the heart. The left ventricular end-systolic volume was computed from the left ventricular end-diastolic volume and the stroke volume. A similar approach was chosen for the right ventricle and the pulmonary circulation. Moreover, the sampling procedure was constrained by a set of well-defined consistency rules to achieve physiological plausibility [455], e.g., left and right ventricular similar stroke volume, physiological pulse pressure, etc. The WBC model was then employed for computing the personalized output measures of interest, thus creating a dataset of training examples.

Two distinct neural networks were defined, to predict time-independent and time-dependent measures of interest respectively. Following the standard approaches, 8000 randomly selected data samples were used for training while the remaining 2000 are used for testing [454]. All input parameters were used as features for the network, leading to an input feature vector of 9 floating-point values. For more stable and faster training, features were rescaled to have the properties of a standard normal distribution with mean 0 and standard deviation of 1.

A schematic representation of the overall modeling method is shown in Fig. 6.12: the deep neural networks are trained offline on the synthetically generated database, and are then applied to predict measures of interest as a function of patient-specific clinical data.

6.3.3 Results and discussion

To assess the model performance, the mean absolute percentage error, which is typically employed in regression tasks evaluation, is computed (see Table 6.4). Pearson correlation between predictions and ground-truth was also considered to assess the fidelity of the AI model. (See Fig. 6.13.)

Important information for the patient condition can be retrieved by assessing the evolution in time of diverse quantities within a heart cycle. One relevant example is the non-invasive

Table 6.4 Results of the deep neural network for time-independent quantities on the test dataset.

Circulation	Parameters	MAPE [%]	Pearson correlation
Systemic	Dead volume	3.87	0.9997
	Time at max. elastance	0.12	0.9995
	Resistance	0.13	0.9999
	Compliance	2.83	0.9895
Pulmonary	Dead volume	13.48	0.9994
	Time at max. elastance	0.10	0.9995
	Resistance	0.26	0.9997
	Compliance	0.76	0.9985

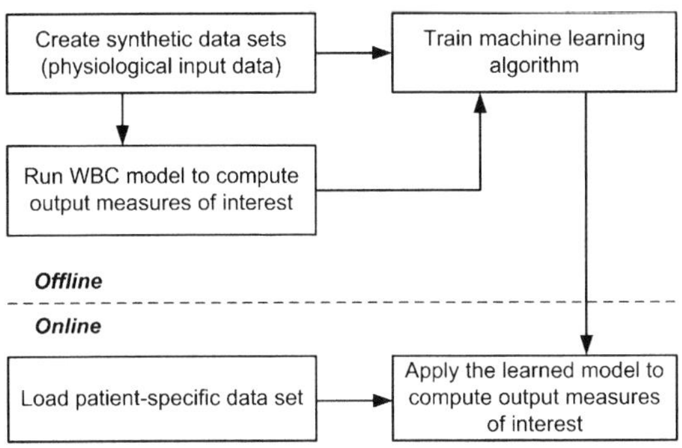

Figure 6.12. Overall workflow of the proposed deep learning based model.

real-time computation of pressure-volume loops. The left ventricular pressure-volume (PV) loop (see examples in Fig. 6.14L) represents an efficient tool for understanding and characterizing cardiac function. It contains information regarding stroke volume, cardiac output, ejection fraction, myocardial contractility, cardiac oxygen consumption, and other important measures of the heart and the systemic circulation. For example, the extent of ventricular remodeling, the degree of ventricular-arterial mismatching [456], and the left ventricular end-diastolic pressure-volume relationship [457] represent strong predictors of congestive heart failure. Pathologies such as left ventricular hypertrophy, dilated cardiomyopathy, aortic and mitral valve stenosis, and regurgitation [137] are manifested in the PV-loop. (See Table 6.5.)

Table 6.5 Results of the Deep Neural Network for time-dependent quantities on the test dataset.

Circulation	Parameters	MAE	RMSE
Systemic	Aortic pressure [mmHg]	1.46	0.35
	Ventricular pressure [mmHg]	1.33	0.18
	Atrial pressure [mmHg]	0.52	0.02
	Ventricular volume [ml]	1.85	0.52
	Aortic Flow rate [ml/s]	6.22	0.08
Pulmonary	Pulmonary artery pressure [mmHg]	0.56	0.06
	Ventricular pressure [mmHg]	0.57	0.01
	Atrial pressure [mmHg]	0.52	0.01
	Ventricular volume [ml]	1.33	0.54
	Pulmonary artery Flow rate [ml/s]	3.13	0.07

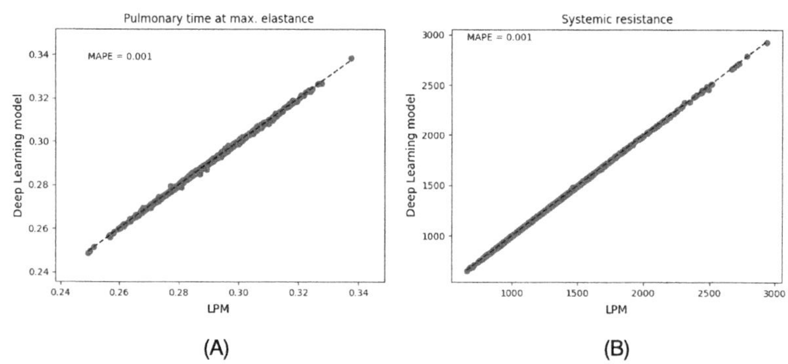

Figure 6.13. Correlation between predictions and ground-truth. (A) Predicted vs. Ground-truth time at max. elastance; (B) Predicted vs. Ground-truth resistance.

The relatively large runtime required for personalizing the whole body circulation model could represent an important drawback for the clinical adoption of closed-loop hemodynamic models in contexts in which their contribution could be relevant, such as for continuous monitoring of patients in intensive care units. Hence, a method for efficient estimation of clinically relevant measures of interest such as the PV loop could represent a powerful diagnostic tool in specific clinical workflows. This work demonstrates the definition of a new model based on two deep neural networks trained on a synthetically generated database of clinically observable quantities (as input) and clinically relevant measures of interest (as ground truth values) estimated by a WBC

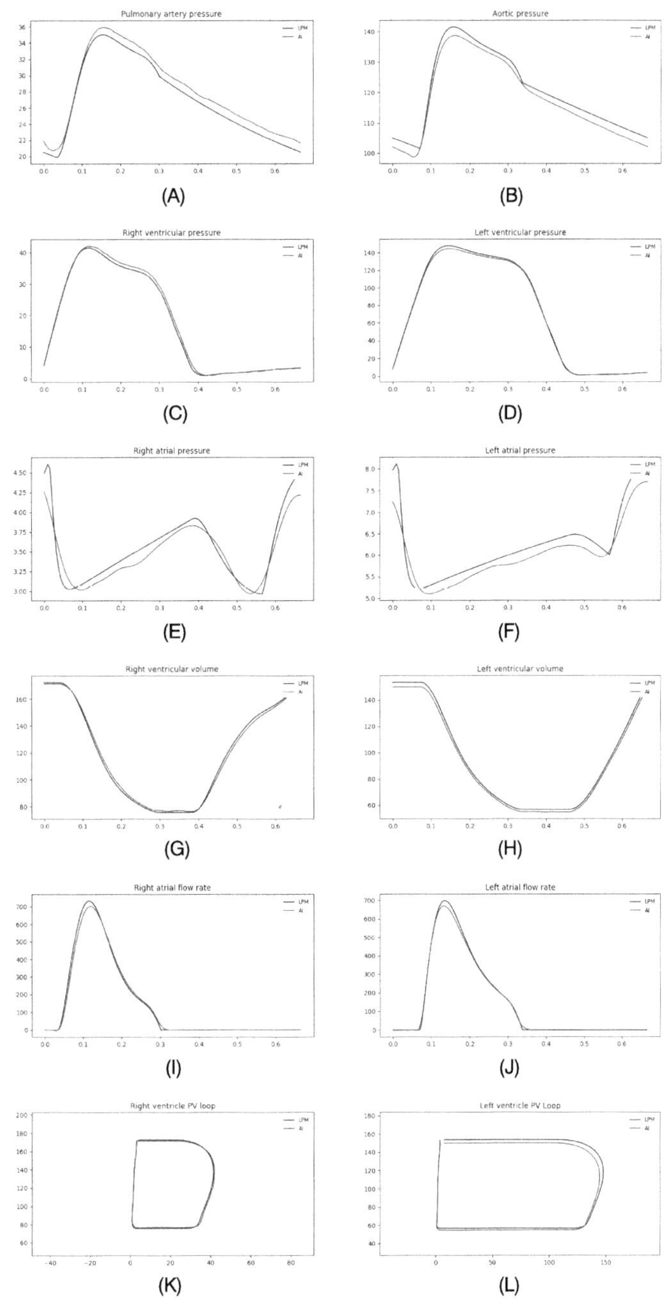

Figure 6.14. Time series predictions (red (mid gray in print version)) vs. ground-truth (blue (dark gray in print version)). (A) Pulmonary artery pressure; (B) Aortic pressure; (C) Right ventricular pressure; (D) Left ventricular pressure; (E) Right atrial pressure; (F) Left atrial pressure; (G) Right ventricular volume; (H) Left ventricular volume; (I) Pulmonary artery flow rate; (J) Aortic flow rate; (K) Right ventricle PV loop; (L) Left ventricle PV loop.

model. The performance is statistically equivalent to that of the computational hemodynamics model, while the inference time is reduced to milliseconds.

It is important to note that the accuracy of the proposed method compared with invasive measurements of quantities of interest depends on the accuracy of the computational model used to generate the training samples. Validation of the method could be based on the extensive validation of the WBC model on patient-specific datasets with invasive measurements; followed by re-training of the neural networks on a newly generated training database.

6.4 Summary

This section presented three clinical applications of AI and computational models. First, we showed how a virtual heart model could be used to plan and guide cardiac resynchronization therapy. Then, AI based alternatives to computational hemodynamics models were presented for decision making in the cathlab.

These AI-enabled modeling solutions can be extended to other applications and opportunities are possible: some natural extensions of use for the virtual heart model are for instance the planning and guidance of ablation therapy for complex arrhythmias, planning of valve surgery, etc. Indeed, the approaches presented in this book could extend to the modeling of other organs and biological systems.

While many challenges still need to be addressed (e.g. noise in the data and uncertainty quantification, validation studies, among others), this technology already shows great potential towards the future of medicine.

Bibliography

1. M. Grieves, Digital Twin: Manufacturing Excellence Through Virtual Factory Replication, White paper, 2014, pp. 1–7.
2. E. Glaessgen, D. Stargel, The digital twin paradigm for future NASA and us air force vehicles, in: 53rd AIAA/ASME/ASCE/AHS/ASC Structures, Structural Dynamics and Materials Conference 20th AIAA/ASME/AHS Adaptive Structures Conference 14th AIAA, 2012, p. 1818.
3. M. Andrychowicz, B. Baker, M. Chociej, R. Jozefowicz, B. McGrew, J. Pachocki, A. Petron, M. Plappert, G. Powell, A. Ray, et al., Learning dexterous in-hand manipulation, arXiv preprint, arXiv:1808.00177, 2018.
4. D. Comaniciu, K. Engel, B. Georgescu, T. Mansi, Shaping the Future Through Innovations: From Medical Imaging to Precision Medicine, 2016.
5. N.A. Trayanova, P.M. Boyle, P.P. Nikolov, Personalized imaging and modeling strategies for arrhythmia prevention and therapy, Current Opinion in Biomedical Engineering 5 (2018) 21–28.
6. E. Kayvanpour, T. Mansi, F. Sedaghat-Hamedani, A. Amr, D. Neumann, B. Georgescu, P. Seegerer, A. Kamen, J. Haas, K.S. Frese, M. Irawati, E. Wirsz, V. King, S. Buss, D. Mereles, E. Zitron, A. Keller, H.A. Katus, D. Comaniciu, B. Meder, Towards personalized cardiology: multi-scale modeling of the failing heart, PLoS ONE 10 (7) (2015) 1–18.
7. P.J. Hunter, T.K. Borg, Integration from proteins to organs: the physiome project, Nature Reviews. Molecular Cell Biology 4 (3) (2003) 237.
8. H. Ashikaga, H. Arevalo, F. Vadakkumpadan, R.C. Blake III, J.D. Bayer, S. Nazarian, M.M. Zviman, H. Tandri, R.D. Berger, H. Calkins, et al., Feasibility of image-based simulation to estimate ablation target in human ventricular arrhythmia, Heart Rhythm 10 (8) (2013) 1109–1116.
9. A.W. Lee, C.M. Costa, M. Strocchi, C.A. Rinaldi, S.A. Niederer, Computational modeling for cardiac resynchronization therapy, Journal of Cardiovascular Translational Research 11 (2) (2018) 92–108.
10. M. Tegmark, Life 3.0: Being Human in the Age of Artificial Intelligence, Knopf, 2017.
11. D. Silver, A. Huang, C.J. Maddison, A. Guez, L. Sifre, G. Van Den Driessche, J. Schrittwieser, I. Antonoglou, V. Panneershelvam, M. Lanctot, et al., Mastering the game of go with deep neural networks and tree search, Nature 529 (7587) (2016) 484.
12. A. Hodgkin, A. Huxley, A quantitative description of ion currents and its applications to conduction and excitation in nerve membranes, Journal of Physiology 117 (4) (1952) 500–544.
13. K. Ten Tusscher, D. Noble, P.-J. Noble, A.V. Panfilov, A model for human ventricular tissue, American Journal of Physiology. Heart and Circulatory Physiology 286 (4) (2004) H1573–H1589.
14. J.J. Rice, P.P. De Tombe, Approaches to modeling crossbridges and calcium-dependent activation in cardiac muscle, Progress in Biophysics and Molecular Biology 85 (2–3) (2004) 179–195.
15. D.M. Bers, Cardiac excitation–contraction coupling, Nature 415 (6868) (2002) 198.
16. D.A. Beard, Modeling of oxygen transport and cellular energetics explains observations on in vivo cardiac energy metabolism, PLoS Computational Biology 2 (9) (2006) e107.
17. C.M. Lloyd, J.R. Lawson, P.J. Hunter, P.F. Nielsen, The cellml model repository, Bioinformatics 24 (18) (2008) 2122–2123.
18. M. Sermesant, R. Chabiniok, P. Chinchapatnam, T. Mansi, F. Billet, P. Moireau, J.-M. Peyrat, K. Wong, J. Relan, K. Rhode, et al., Patient-specific electromechanical models of the heart for the prediction of pacing acute effects in crt: a preliminary clinical validation, Medical Image Analysis 16 (1) (2012) 201–215.
19. A. Prakosa, H.J. Arevalo, D. Deng, P.M. Boyle, P.P. Nikolov, H. Ashikaga, J.J. Blauer, E. Ghafoori, C.J. Park, R.C. Blake, et al., Personalized virtual-heart technology for guiding the ablation of infarct-related ventricular tachycardia, Nature Biomedical Engineering 2 (10) (2018) 732.
20. N. Cedilnik, J. Duchateau, R. Dubois, F. Sacher, P. Jaïs, H. Cochet, M. Sermesant, Fast personalized electrophysiological models from computed tomography images for ventricular tachycardia ablation planning, EP Europace 20 (suppl 3) (2018), iii94–iii101.
21. P.M. Boyle, T. Zghaib, S. Zahid, R.L. Ali, D. Deng, W.H. Franceschi, J.B. Hakim, M.J. Murphy, A. Prakosa, S.L. Zimmerman, et al., Computationally guided personalized targeted ablation of persistent atrial fibrillation, Nature Biomedical Engineering (2019) 1–10.

22. T. Mansi, I. Voigt, B. Georgescu, X. Zheng, E.A. Mengue, M. Hackl, R.I. Ionasec, T. Noack, J. Seeburger, D. Comaniciu, An integrated framework for finite-element modeling of mitral valve biomechanics from medical images: application to mitralclip intervention planning, Medical Image Analysis 16 (7) (2012) 1330–1346.
23. A. Drach, A.H. Khalighi, M.S. Sacks, A comprehensive pipeline for multi-resolution modeling of the mitral valve: Validation, computational efficiency, and predictive capability, International Journal for Numerical Methods in Biomedical Engineering 34 (2) (2018) e2921.
24. T. Mansi, B. André, M. Lynch, M. Sermesant, H. Delingette, Y. Boudjemline, N. Ayache, Virtual pulmonary valve replacement interventions with a personalised cardiac electromechanical model, in: Recent Advances in the 3D Physiological Human, Springer, 2009, pp. 75–90.
25. T.M. van Bakel, K.D. Lau, J. Hirsch-Romano, S. Trimarchi, A.L. Dorfman, C.A. Figueroa, Patient-specific modeling of hemodynamics: supporting surgical planning in a fontan circulation correction, Journal of Cardiovascular Translational Research 11 (2) (2018) 145–155.
26. D. Neumann, T. Mansi, B. Georgescu, A. Kamen, E. Kayvanpour, A. Amr, F. Sedaghat-Hamedani, J. Haas, H. Katus, B. Meder, J. Hornegger, D. Comaniciu, Robust image-based estimation of cardiac tissue parameters and their uncertainty from noisy data, in: International Conference on Medical Image Computing and Computer-Assisted Intervention, in: LNCS, vol. 8674, Springer, 2014, pp. 9–16.
27. J. Dhamala, H.J. Arevalo, J. Sapp, B.M. Horácek, K.C. Wu, N.A. Trayanova, L. Wang, Quantifying the uncertainty in model parameters using Gaussian process-based Markov chain Monte Carlo in cardiac electrophysiology, Medical Image Analysis 48 (2018) 43–57.
28. J. McCarthy, Artificial intelligence, logic and formalizing common sense, in: Philosophical Logic and Artificial Intelligence, Springer, 1989, pp. 161–190.
29. W.S. McCulloch, W. Pitts, A logical calculus of the ideas immanent in nervous activity, The Bulletin of Mathematical Biophysics 5 (4) (1943) 115–133.
30. A. Krizhevsky, I. Sutskever, H. geoffrey e., "alex net, ", Advances in Neural Information Processing Systems 25 (2012) 1–9.
31. Y. Zheng, A. Barbu, B. Georgescu, M. Scheuering, D. Comaniciu, Four-chamber heart modeling and automatic segmentation for 3-d cardiac ct volumes using marginal space learning and steerable features, IEEE Transactions on Medical Imaging 27 (11) (2008) 1668–1681.
32. R.I. Ionasec, I. Voigt, B. Georgescu, Y. Wang, H. Houle, F. Vega-Higuera, N. Navab, D. Comaniciu, Patient-specific modeling and quantification of the aortic and mitral valves from 4-D cardiac CT and TEE, IEEE Transactions on Medical Imaging 29 (Sept. 2010) 1636–1651.
33. K. Hammernik, T. Klatzer, E. Kobler, M.P. Recht, D.K. Sodickson, T. Pock, F. Knoll, Learning a variational network for reconstruction of accelerated mri data, Magnetic Resonance in Medicine 79 (6) (2018) 3055–3071.
34. R. Booij, R.P. Budde, M.L. Dijkshoorn, M. van Straten, Accuracy of automated patient positioning in ct using a 3d camera for body contour detection, European Radiology 29 (4) (2019) 2079–2088.
35. V. Singh, K. Ma, B. Tamersoy, Y.-J. Chang, A. Wimmer, T. O'Donnell, T. Chen, Darwin: deformable patient avatar representation with deep image network, in: International Conference on Medical Image Computing and Computer-Assisted Intervention, Springer, 2017, pp. 497–504.
36. B. Lou, S. Doken, T. Zhuang, D. Wingerter, M. Gidwani, N. Mistry, L. Ladic, A. Kamen, M.E. Abazeed, An image-based deep learning framework for individualising radiotherapy dose: a retrospective analysis of outcome prediction, The Lancet Digital Health 1 (3) (2019) e136–e147.
37. T. Mansi, B. Georgescu, J. Hussan, P.J. Hunter, A. Kamen, D. Comaniciu, Data-driven reduction of a cardiac myofilament model, in: International Conference on Functional Imaging and Modeling of the Heart, Springer, 2013, pp. 232–240.
38. S. He, Y. Li, Y. Feng, S. Ho, S. Ravanbakhsh, W. Chen, B. Póczos, Learning to predict the cosmological structure formation, Proceedings of the National Academy of Sciences (2019) 201821458.
39. B. Kim, V.C. Azevedo, N. Thuerey, T. Kim, M. Gross, B. Solenthaler, Deep Fluids: A Generative Network for Parameterized Fluid Simulations, Computer Graphics Forum, vol. 38, Wiley Online Library, 2019, pp. 59–70.
40. R. Modrzejewski, T. Collins, A. Bartoli, A. Hostettler, J. Marescaux, Soft-body registration of pre-operative 3d models to intra-operative rgbd partial body scans, in: International Conference on Medical Image Computing and Computer-Assisted Intervention, Springer, 2018, pp. 39–46.

41. P.M. DeVries, F. Viégas, M. Wattenberg, B.J. Meade, Deep learning of aftershock patterns following large earthquakes, Nature 560 (7720) (2018) 632.
42. A.M. Katz, Physiology of the Heart, Lippincott Williams & Wilkins, 2010.
43. M.P. Nash, P.J. Hunter, Computational mechanics of the heart, Journal of Elasticity and the Physical Science of Solids 61 (1–3) (2000) 113–141.
44. J.M. Guccione, G.S. Kassab, M.B. Ratcliffe, et al., Computational Cardiovascular Mechanics, Springer, 2010.
45. M. Sermesant, H. Delingette, N. Ayache, An electromechanical model of the heart for image analysis and simulation, IEEE Transactions on Medical Imaging 25 (5) (2006) 612–625.
46. V.Y. Wang, H. Lam, D.B. Ennis, B.R. Cowan, A.A. Young, M.P. Nash, Modelling passive diastolic mechanics with quantitative mri of cardiac structure and function, Medical Image Analysis 13 (5) (2009) 773–784.
47. R. Chabiniok, V.Y. Wang, M. Hadjicharalambous, L. Asner, J. Lee, M. Sermesant, E. Kuhl, A.A. Young, P. Moireau, M.P. Nash, et al., Multiphysics and multiscale modelling, data–model fusion and integration of organ physiology in the clinic: ventricular cardiac mechanics, Interface Focus 6 (2) (2016) 20150083.
48. K. Kunzelman, D.R. Einstein, R. Cochran, Fluid-structure interaction models of the mitral valve: function in normal and pathological states, Philosophical Transactions of the Royal Society of London. Series B, Biological Sciences 362 (1484) (2007) 1393–1406.
49. D.D. Streeter Jr, H.M. Spotnitz, D.P. Patel, J. Ross Jr, E.H. Sonnenblick, Fiber orientation in the canine left ventricle during diastole and systole, Circulation Research 24 (3) (1969) 339–347.
50. F.J. Vetter, A.D. McCulloch, Three-dimensional analysis of regional cardiac function: a model of rabbit ventricular anatomy, Progress in Biophysics and Molecular Biology 69 (2–3) (1998) 157–183.
51. M. Pop, M. Sermesant, T. Mansi, E. Crystal, S. Ghate, J.-M. Peyrat, I. Lashevsky, B. Qiang, E. McVeigh, N. Ayache, et al., Correspondence between simple 3-d mri-based computer models and in-vivo ep measurements in swine with chronic infarctions, IEEE Transactions on Biomedical Engineering 58 (12) (2011) 3483–3486.
52. S. Nielles-Vallespin, C. Mekkaoui, P. Gatehouse, T.G. Reese, J. Keegan, P.F. Ferreira, S. Collins, P. Speier, T. Feiweier, R. de Silva, et al., In vivo diffusion tensor mri of the human heart: reproducibility of breath-hold and navigator-based approaches, Magnetic Resonance in Medicine 70 (2) (2013) 454–465.
53. J.-M. Peyrat, M. Sermesant, X. Pennec, H. Delingette, C. Xu, E. McVeigh, N. Ayache, Towards a statistical atlas of cardiac fiber structure, in: International Conference on Medical Image Computing and Computer-Assisted Intervention, Springer, 2006, pp. 297–304.
54. H. Lombaert, J.-M. Peyrat, P. Croisille, S. Rapacchi, L. Fanton, F. Cheriet, P. Clarysse, I. Magnin, H. Delingette, N. Ayache, Human atlas of the cardiac fiber architecture: study on a healthy population, IEEE Transactions on Medical Imaging 31 (7) (2012) 1436–1447.
55. T. Arts, R.S. Reneman, P.C. Veenstra, A model of the mechanics of the left ventricle, Annals of Biomedical Engineering 7 (3–4) (1979) 299–318.
56. I.J. LeGrice, B. Smaill, L. Chai, S. Edgar, J. Gavin, P.J. Hunter, Laminar structure of the heart: ventricular myocyte arrangement and connective tissue architecture in the dog, American Journal of Physiology. Heart and Circulatory Physiology 269 (2) (1995) H571–H582.
57. J.D. Bayer, R.C. Blake, G. Plank, N.A. Trayanova, A novel rule-based algorithm for assigning myocardial fiber orientation to computational heart models, Annals of Biomedical Engineering 40 (10) (2012) 2243–2254.
58. J.L. Titus, Normal anatomy of the human cardiac conduction system, Anesthesia and Analgesia 52 (4) (1973) 508–514.
59. J. Bigger, W. Mandel, Effect of lidocaine on conduction in canine Purkinje fibers and at the ventricular muscle–Purkinje fiber junction, The Journal of Pharmacology and Experimental Therapeutics 172 (1970) 239–254.
60. D. Durrer, R.T. Van Dam, G. Freud, M. Janse, F. Meijler, R. Arzbaecher, Total excitation of the isolated human heart, Circulation 41 (6) (1970) 899–912.
61. D.A. Rawling, R.W. Joyner, E.D. Overholt, Variations in the functional electrical coupling between the subendocardial Purkinje and ventricular layers of the canine left ventricle, Circulation Research 57 (2) (1985) 252–261.
62. R.J. Prineas, R.S. Crow, Z.-M. Zhang, The Minnesota Code Manual of Electrocardiographic Findings, Springer Science & Business Media, 2009.
63. M. Shenasa, G. Hindricks, D.J. Callans, J.M. Miller, M.E. Josephson, Cardiac Mapping, Wiley Online Library, 2003.
64. R. FitzHugh, Impulses and physiological states in theoretical models of nerve membrane, Biophysical Journal 1 (1961) 445–466.

65. J. Nagumo, S. Arimoto, S. Yoshizawa, An active pulse transmission line simulating nerve axon, Proceeding IRE 50 (1962) 2061–2070.
66. R. Clayton, A. Panfilov, A guide to modelling cardiac electrical activity in anatomically detailed ventricles, Progress in Biophysics and Molecular Biology 96 (1–3) (2008) 19–43.
67. D. Noble, A modification of the Hodgkin–Huxley equations applicable to Purkinje fibre action and pacemaker potentials, The Journal of Physiology 160 (2) (1962) 317–352.
68. C.-h. Luo, Y. Rudy, A model of the ventricular cardiac action potential. Depolarization, repolarization, and their interaction, Circulation Research 68 (6) (1991) 1501–1526.
69. D. Noble, A. Varghese, P. Kohl, P. Noble, Improved Guinea-pig ventricular cell model incorporating a diadic space, ikr and iks, and length-and tension-dependent processes, The Canadian Journal of Cardiology 14 (1) (1998) 123–134.
70. K.M. Holzem, E.J. Madden, I.R. Efimov, Human cardiac systems electrophysiology and arrhythmogenesis: iteration of experiment and computation, Europace 16 (suppl 4) (2014), iv77–iv85.
71. C.C. Mitchell, D.G. Schaeffer, A two-current model for the dynamics of cardiac membrane, Bulletin of Mathematical Biology 65 (5) (2003) 767–793.
72. C. Corrado, J.-F. Gerbeau, P. Moireau, Identification of weakly coupled multiphysics problems. application to the inverse problem of electrocardiography, Journal of Computational Physics 283 (2015) 271–298.
73. J. Relan, M. Sermesant, M. Pop, H. Delingette, M. Sorine, G.A. Wright, N. Ayache, Parameter estimation of a 3d cardiac electrophysiology model including the restitution curve using optical and mr data, in: World Congress on Medical Physics and Biomedical Engineering, September 7–12, 2009, Munich, Germany, Springer, 2009, pp. 1716–1719.
74. H. Talbot, S. Cotin, R. Razavi, C. Rinaldi, H. Delingette, Personalization of Cardiac Electrophysiology Model Using the Unscented Kalman Filtering, 2015.
75. Y. Coudière, C. Pierre, Stability and convergence of a finite volume method for two systems of reaction–diffusion equations in electro-cardiology, Nonlinear Analysis: Real World Applications 7 (4) (2006) 916–935.
76. Y. Bourgault, Y. Coudiere, C. Pierre, Existence and uniqueness of the solution for the bidomain model used in cardiac electrophysiology, Nonlinear Analysis: Real World Applications 10 (1) (2009) 458–482.
77. P.C. Franzone, L. Guerri, S. Rovida, Wavefront propagation in an activation model of the anisotropic cardiac tissue: asymptotic analysis and numerical simulations, Journal of Mathematical Biology 28 (2) (1990) 121–176.
78. J.P. Keener, J. Sneyd, Mathematical Physiology, vol. 1, Springer, 1998.
79. M. Sermesant, E. Konukoglu, H. Delingette, Y. Coudière, P. Chinchapatnam, K.S. Rhode, R. Razavi, N. Ayache, An anisotropic multi-front fast marching method for real-time simulation of cardiac electrophysiology, in: International Conference on Functional Imaging and Modeling of the Heart, Springer, 2007, pp. 160–169.
80. J.A. Sethian, Level Set Methods and Fast Marching Methods: Evolving Interfaces in Computational Geometry Fluid Mechanics, Computer Vision, and Materials Science, vol. 3, Cambridge University Press, 1999.
81. E. Konukoglu, M. Sermesant, O. Clatz, J.-M. Peyrat, H. Delingette, N. Ayache, A recursive anisotropic fast marching approach to reaction diffusion equation: application to tumor growth modeling, in: Biennial International Conference on Information Processing in Medical Imaging, Springer, 2007, pp. 687–699.
82. M. Wallman, N.P. Smith, B. Rodriguez, Computational methods to reduce uncertainty in the estimation of cardiac conduction properties from electroanatomical recordings, Medical Image Analysis 18 (1) (2014) 228–240.
83. E. Pernod, M. Sermesant, E. Konukoglu, J. Relan, H. Delingette, N. Ayache, A multi-front Eikonal model of cardiac electrophysiology for interactive simulation of radio-frequency ablation, Computers & Graphics 35 (2) (2011) 431–440.
84. K.H. Tusscher, A.V. Panfilov, Modelling of the ventricular conduction system, Progress in Biophysics and Molecular Biology 96 (1–3) (2008) 152–170.
85. T. Ijiri, T. Ashihara, T. Yamaguchi, K. Takayama, T. Igarashi, T. Shimada, T. Namba, R. Haraguchi, K. Nakazawa, A procedural method for modeling the Purkinje fibers of the heart, Journal of Physiological Sciences 58 (7) (2008) 481–486.
86. R. Sebastian, V. Zimmerman, D. Romero, A.F. Frangi, Construction of a computational anatomical model of the peripheral cardiac conduction system, IEEE Transactions on Biomedical Engineering 58 (12) (2011) 3479–3482.
87. E.J. Vigmond, C. Clements, D.M. McQueen, C.S. Peskin, Effect of bundle branch block on cardiac output: a whole heart simulation study, Progress in Biophysics and Molecular Biology 97 (2–3) (2008) 520–542.

88. P. Colli-Franzone, L. Guerri, M. Pennachio, B. Taccardi, Spread of excitation in 3-d models of the anisotropic cardiac tissue. ii. Effects of fiber architecture and ventricular geometry, Mathematical Biosciences 147 (1998) 131–171.
89. R.C. Kerckoffs, O.P. Faris, P.H. Bovendeerd, F.W. Prinzen, K. Smits, E.R. McVeigh, T. Arts, Timing of depolarization and contraction in the paced canine left ventricle: model and experiment, Journal of Cardiovascular Electrophysiology 14 (2003) S188–S195.
90. F.N. Wilson, A.G. Macleod, P.S. Barker, The potential variations produced by the heart beat at the apices of einthoven's triangle, American Heart Journal 7 (2) (1931) 207–211.
91. F.N. Wilson, F.D. Johnston, F.F. Rosenbaum, H. Erlanger, C.E. Kossmann, H. Hecht, N. Cotrim, R.M. de Oliveira, R. Scarsi, P.S. Barker, The precordial electrocardiogram, American Heart Journal 27 (1) (1944) 19–85.
92. A. Lyon, A. Mincholé, J.P. Martínez, P. Laguna, B. Rodriguez, Computational techniques for ecg analysis and interpretation in light of their contribution to medical advances, Journal of the Royal Society Interface 15 (138) (2018) 20170821.
93. D.M. Mirvis, Body Surface Electrocardiographic Mapping, vol. 82, Springer Science & Business Media, 2012.
94. Y. Rudy, The electrocardiogram and its relationship to excitation of the heart, Developments in Cardiovascular Medicine (1995), 151:201.
95. B. Messinger-Rapport, Y. Rudy, Noninvasive recovery of epicardial potentials in a realistic heart-torso geometry. Normal sinus rhythm, Circulation Research 66 (4) (1990) 1023–1039.
96. P. Colli-Franzone, L. Guerri, S. Tentoni, C. Viganotti, S. Baruffi, S. Spaggiari, B. Taccardi, A mathematical procedure for solving the inverse potential problem of electrocardiography. Analysis of the time-space accuracy from in vitro experimental data, Mathematical Biosciences 77 (1–2) (1985) 353–396.
97. R.C. Barr, M. Spach, Inverse calculation of qrs-t epicardial potentials from body surface potential distributions for normal and ectopic beats in the intact dog, Circulation Research 42 (5) (1978) 661–675.
98. J. Duchateau, F. Sacher, T. Pambrun, N. Derval, J. Chamorro-Servent, A. Denis, S. Ploux, M. Hocini, P. Jaïs, O. Bernus, et al., Performance and limitations of noninvasive cardiac activation mapping, Heart Rhythm 16 (3) (2019) 435–442.
99. M. Chhay, Y. Coudière, R. Turpault, How to Compute the Extracellular Potential in Electrocardiology From an Extended Monodomain Model, 2012.
100. P.J. Hunter, A.J. Pullan, B.H. Smaill, Modeling total heart function, Annual Review of Biomedical Engineering 5 (1) (2003) 147–177.
101. Y.-c. Fung, Biomechanics: Mechanical Properties of Living Tissues, Springer Science & Business Media, 2013.
102. S. Göktepe, A. Menzel, E. Kuhl, The generalized hill model: a kinematic approach towards active muscle contraction, Journal of the Mechanics and Physics of Solids 72 (2014) 20–39.
103. D. Chapelle, P. Le Tallec, P. Moireau, M. Sorine, Energy-preserving muscle tissue model: formulation and compatible discretizations, International Journal of Multiscale Computational Engineering 10 (2) (2012).
104. K.-J. Bathe, Finite Element Procedures, Klaus-Jurgen Bathe, 2006.
105. S. Marchesseau, T. Heimann, S. Chatelin, R. Willinger, H. Delingette, Fast porous visco-hyperelastic soft tissue model for surgery simulation: application to liver surgery, Progress in Biophysics and Molecular Biology 103 (2–3) (2010) 185–196.
106. K. Miller, G. Joldes, D. Lance, A. Wittek, Total Lagrangian explicit dynamics finite element algorithm for computing soft tissue deformation, Communications in Numerical Methods in Engineering 23 (2) (2007) 121–134.
107. S.A. Niederer, J. Lumens, N.A. Trayanova, Computational models in cardiology, Nature Reviews Cardiology (2018) 1.
108. W. Kroon, T. Delhaas, T. Arts, P. Bovendeerd, Computational modeling of volumetric soft tissue growth: application to the cardiac left ventricle, Biomechanics and Modeling in Mechanobiology 8 (4) (2009) 301–309.
109. S. Göktepe, O.J. Abilez, K.K. Parker, E. Kuhl, A multiscale model for eccentric and concentric cardiac growth through sarcomerogenesis, Journal of Theoretical Biology 265 (3) (2010) 433–442.
110. R.C. Kerckhoffs, J.H. Omens, A.D. McCulloch, A single strain-based growth law predicts concentric and eccentric cardiac growth during pressure and volume overload, Mechanics Research Communications 42 (2012) 40–50.
111. N. Cohen, V.S. Deshpande, J.W. Holmes, R.M. McMeeking, A microscopically motivated model for the remodeling of cardiomyocytes, Biomechanics and Modeling in Mechanobiology (2019) 1–13.

112. S. Dokos, B.H. Smaill, A.A. Young, I.J. LeGrice, Shear properties of passive ventricular myocardium, American Journal of Physiology. Heart and Circulatory Physiology 283 (6) (2002) H2650–H2659.
113. S. Göktepe, S. Acharya, J. Wong, E. Kuhl, Computational modeling of passive myocardium, International Journal for Numerical Methods in Biomedical Engineering 27 (1) (2011) 1–12.
114. J. Guccione, A. McCulloch, Mechanics of active contraction in cardiac muscle: part I—constitutive relations for fiber stress that describe deactivation, Journal of Biomechanical Engineering 115 (1) (1993) 72–81.
115. S. Land, V. Gurev, S. Arens, C.M. Augustin, L. Baron, R. Blake, C. Bradley, S. Castro, A. Crozier, M. Favino, et al., Verification of cardiac mechanics software: benchmark problems and solutions for testing active and passive material behaviour, Proceedings of the Royal Society A. Mathematical, Physical and Engineering Sciences 471 (2184) (2015) 20150641.
116. P.J. Hunter, B.H. Smaill, The analysis of cardiac function: a continuum approach, Progress in Biophysics and Molecular Biology 52 (2) (1988) 101–164.
117. S.A. Niederer, N.P. Smith, An improved numerical method for strong coupling of excitation and contraction models in the heart, Progress in Biophysics and Molecular Biology 96 (1–3) (2008) 90–111.
118. K.D. Costa, J.W. Holmes, A.D. McCulloch, Modelling cardiac mechanical properties in three dimensions, Philosophical Transactions of the Royal Society of London. Series A: Mathematical, Physical and Engineering Sciences 359 (1783) (2001) 1233–1250.
119. H. Schmid, M. Nash, A. Young, P. Hunter, Myocardial material parameter estimation—a comparative study for simple shear, Journal of Biomechanical Engineering 128 (5) (2006) 742–750.
120. G.A. Holzapfel, R.W. Ogden, Constitutive modelling of passive myocardium: a structurally based framework for material characterization, Philosophical Transactions of the Royal Society of London A: Mathematical, Physical and Engineering Sciences 367 (1902) (2009) 3445–3475.
121. S. Marchesseau, H. Delingette, M. Sermesant, R. Cabrera-Lozoya, C. Tobon-Gomez, P. Moireau, R.F. i Ventura, K. Lekadir, A. Hernandez, M. Garreau, et al., Personalization of a cardiac electromechanical model using reduced order unscented Kalman filtering from regional volumes, Medical Image Analysis 17 (7) (2013) 816–829.
122. Y. Aboelkassem, K.J. McCabe, G. Huber, M. Regnier, J.A. McCammon, A.D. McCulloch, A stochastic coarse-graining multiscale model of cardiac thin filament activation using Brownian and Langevin dynamics, Biophysical Journal (2019).
123. J.J. Rice, F. Wang, D.M. Bers, P.P. De Tombe, Approximate model of cooperative activation and crossbridge cycling in cardiac muscle using ordinary differential equations, Biophysical Journal 95 (5) (2008) 2368–2390.
124. K. Tran, N.P. Smith, D.S. Loiselle, E.J. Crampin, A metabolite-sensitive, thermodynamically constrained model of cardiac cross-bridge cycling: implications for force development during ischemia, Biophysical Journal 98 (2) (2010) 267–276.
125. N.A. Trayanova, Whole-heart modeling: applications to cardiac electrophysiology and electromechanics, Circulation Research 108 (1) (2011) 113–128.
126. J. Bestel, F. Clément, M. Sorine, A biomechanical model of muscle contraction, in: International Conference on Medical Image Computing and Computer-Assisted Intervention, Springer, 2001, pp. 1159–1161.
127. T. Arts, P. Bovendeerd, F.W. Prinzen, R.S. Reneman, Relation between left ventricular cavity pressure and volume and systolic fiber stress and strain in the wall, Biophysical Journal 59 (1) (1991) 93–102.
128. M. Caruel, R. Chabiniok, P. Moireau, Y. Lecarpentier, D. Chapelle, Dimensional reductions of a cardiac model for effective validation and calibration, Biomechanics and Modeling in Mechanobiology 13 (4) (2014) 897–914.
129. R. Molléro, X. Pennec, H. Delingette, A. Garny, N. Ayache, M. Sermesant, Multifidelity-cma: a multifidelity approach for efficient personalisation of 3d cardiac electromechanical models, Biomechanics and Modeling in Mechanobiology 17 (1) (2018) 285–300.
130. D.M. McQueen, C.S. Peskin, A three-dimensional computer model of the human heart for studying cardiac fluid dynamics, ACM Siggraph Computer Graphics 34 (1) (2000) 56–60.
131. V. Mihalef, R.I. Ionasec, P. Sharma, B. Georgescu, I. Voigt, M. Suehling, D. Comaniciu, Patient-specific modelling of whole heart anatomy, dynamics and haemodynamics from four-dimensional cardiac ct images, Interface Focus 1 (3) (2011) 286–296.
132. T. Arts, T. Delhaas, P. Bovendeerd, X. Verbeek, F.W. Prinzen, Adaptation to mechanical load

determines shape and properties of heart and circulation: the circadapt model, American Journal of Physiology. Heart and Circulatory Physiology 288 (4) (2005) H1943–H1954.
133. R.C. Kerckhoffs, M.L. Neal, Q. Gu, J.B. Bassingthwaighte, J.H. Omens, A.D. McCulloch, Coupling of a 3d finite element model of cardiac ventricular mechanics to lumped systems models of the systemic and pulmonic circulation, Annals of Biomedical Engineering 35 (1) (2007) 1–18.
134. S. Niederer, K. Rhode, R. Razavi, N. Smith, The importance of model parameters and boundary conditions in whole organ models of cardiac contraction, in: International Conference on Functional Imaging and Modeling of the Heart, Springer, 2009, pp. 348–356.
135. T. Mansi, Image-Based Physiological and Statistical Models of the Heart: Application to Tetralogy of Fallot, PhD thesis, École Nationale Supérieure des Mines de Paris, 2010.
136. M.R. Pfaller, J.M. Hörmann, M. Weigl, A. Nagler, R. Chabiniok, C. Bertoglio, W.A. Wall, The importance of the pericardium for cardiac biomechanics: from physiology to computational modeling, Biomechanics and Modeling in Mechanobiology 18 (2) (2019) 503–529.
137. J. Hall, Guyton and Hall Textbook of Medical Physiology, 12th ed, Saunders Elsevier, 2011.
138. G. Mancia, G. De Backer, A. Dominiczak, R. Cifkova, R. Fagard, G. Germano, G. Grassi, A.M. Heagerty, S.E. Kjeldsen, S. Laurent, et al., 2007 guidelines for the management of arterial hypertension: the task force for the management of arterial hypertension of the European society of hypertension (esh) and of the European society of cardiology (esc), European Heart Journal 28 (12) (2007) 1462–1536.
139. L.M. Ruilope, R.E. Schmieder, Left ventricular hypertrophy and clinical outcomes in hypertensive patients, American Journal of Hypertension 21 (5) (2008) 500–508.
140. Luca Formaggia, Alfio Quarteroni, Allesandro Veneziani, Cardiovascular Mathematics: Modeling and simulation of the circulatory system, vol. 1, Springer Science & Business Media, 2010.
141. Matteo Nobili, Umberto Morbiducci, Raffaele Ponzini, Costantino Del Gaudio, Antonio Balducci, Mauro Grigioni, Franco Maria Montevecchi, Alberto Redaelli, Numerical simulation of the dynamics of a bileaflet prosthetic heart valve using a fluid–structure interaction approach, Journal of Biomechanics 41 (11) (2008) 2539–2550.
142. F. Viscardi, C. Vergara, L. Antiga, S. Merelli, A. Veneziani, G. Puppini, G. Faggian, A. Mazzucco, G.B. Luciani, Comparative finite element model analysis of ascending aortic flow in bicuspid and tricuspid aortic valve, Artificial Organs 34 (12) (2010) 1114–1120.
143. A. Quaini, S. Canic, G. Guidoboni, R. Glowinski, S.R. Igo, C.J. Hartley, W.A. Zoghbi, S.H. Little, A three-dimensional computational fluid dynamics model of regurgitant mitral valve flow: validation against in vitro standards and 3d color Doppler methods, Cardiovascular Engineering and Technology 2 (2) (2011) 77–89.
144. A. Mitoh, T. Yano, K. Sekine, Y. Mitamura, E. Okamoto, D.-W. Kim, R. Yozu, S. Kawada, Computational fluid dynamics analysis of an intra-cardiac axial flow pump, Artificial Organs 27 (1) (2003) 34–40.
145. Y. Bazilevs, J. Gohean, T. Hughes, R.D. Moser, Y. Zhang, Patient-specific isogeometric fluid-structure interaction analysis of thoracic aortic blood flow due to implantation of the jarvik 2000 left ventricular assist device, Computer Methods in Applied Mechanics and Engineering 198 (45–46) (2009) 3534–3550.
146. K.H. Fraser, M.E. Taskin, B.P. Griffith, Z.J. Wu, The use of computational fluid dynamics in the development of ventricular assist devices, Medical Engineering & Physics 33 (3) (2011) 263–280.
147. A.L. Marsden, A.J. Bernstein, V.M. Reddy, S.C. Shadden, R.L. Spilker, F.P. Chan, C.A. Taylor, J.A. Feinstein, Evaluation of a novel y-shaped extracardiac fontan baffle using computational fluid dynamics, Journal of Thoracic and Cardiovascular Surgery 137 (2) (2009) 394–403.
148. K.K. Whitehead, K. Pekkan, H.D. Kitajima, S.M. Paridon, A.P. Yoganathan, M.A. Fogel, Nonlinear power loss during exercise in single-ventricle patients after the fontan: insights from computational fluid dynamics, Circulation 116 (11 supplement) (2007), I–165.
149. M.R. De Leval, G. Dubini, H. Jalali, R. Pietrabissa, et al., Use of computational fluid dynamics in the design of surgical procedures: application to the study of competitive flows in cavopulmonary connections, The Journal of Thoracic and Cardiovascular Surgery 111 (3) (1996) 502–513.
150. L. Formaggia, D. Lamponi, M. Tuveri, A. Veneziani, Numerical modeling of 1d arterial networks coupled with a lumped parameters description of the heart, Computer Methods in Biomechanics and Biomedical Engineering 9 (5) (2006) 273–288.
151. L.M. Itu, P. Sharma, B. Georgescu, A. Kamen, C. Suciu, D. Comaniciu, Model based non-invasive estimation of pv loop from echocardiography, in: 2014 36th Annual International Conference of the

IEEE Engineering in Medicine and Biology Society, 2014, pp. 6774–6777.
152. A. Quarteroni, A. Veneziani, C. Vergara, Geometric multiscale modeling of the cardiovascular system, between theory and practice, Computer Methods in Applied Mechanics and Engineering 302 (2016) 193–252.
153. A. Noordergraaf, P.D. Verdouw, H.B. Boom, The use of an analog computer in a circulation model, Progress in Cardiovascular Diseases 5 (5) (1963) 419–439.
154. N. Westerhof, F. Bosman, C.J. De Vries, A. Noordergraaf, Analog studies of the human systemic arterial tree, Journal of Biomechanics 2 (2) (1969) 121–143.
155. S. Sherwin, L. Formaggia, J. Peiro, V. Franke, Computational modelling of 1d blood flow with variable mechanical properties and its application to the simulation of wave propagation in the human arterial system, International Journal for Numerical Methods in Fluids 43 (6–7) (2003) 673–700.
156. L. Formaggia, D. Lamponi, A. Quarteroni, One-dimensional models for blood flow in arteries, Journal of Engineering Mathematics 47 (3–4) (2003) 251–276.
157. J. Mynard, P. Nithiarasu, A 1d arterial blood flow model incorporating ventricular pressure, aortic valve and regional coronary flow using the locally conservative Galerkin (lcg) method, Communications in Numerical Methods in Engineering 24 (5) (2008) 367–417.
158. J. Alastruey, A.W. Khir, K.S. Matthys, P. Segers, S.J. Sherwin, P.R. Verdonck, K.H. Parker, J. Peiró, Pulse wave propagation in a model human arterial network: assessment of 1-d visco-elastic simulations against in vitro measurements, Journal of Biomechanics 44 (12) (2011) 2250–2258.
159. L.O. Müller, E.F. Toro, A global multiscale mathematical model for the human circulation with emphasis on the venous system, International Journal for Numerical Methods in Biomedical Engineering 30 (7) (2014) 681–725.
160. P.J. Blanco, S.M. Watanabe, M.A.R. Passos, P.A. Lemos, R.A. Feijóo, An anatomically detailed arterial network model for one-dimensional computational hemodynamics, IEEE Transactions on Biomedical Engineering 62 (2) (2014) 736–753.
161. H. Suga, Theoretical analysis of a left-ventricular pumping model based on the systolic time-varying pressure/volume ratio, IEEE Transactions on Biomedical Engineering 1 (1971) 47–55.
162. S.G. Shroff, J.S. Janicki, K.T. Weber, Evidence and quantitation of left ventricular systolic resistance, The American Journal of Physiology 249 (2 Pt 2) (1985) H358–H370.
163. J. Mynard, M. Davidson, D. Penny, J. Smolich, A simple, versatile valve model for use in lumped parameter and one-dimensional cardiovascular models, International Journal for Numerical Methods in Biomedical Engineering 28 (6–7) (2012) 626–641.
164. N. Westerhof, J.-W. Lankhaar, B.E. Westerhof, The arterial windkessel, Medical & Biological Engineering & Computing 47 (2) (2009) 131–141.
165. T. Wehrum, F. Guenther, A. Fuchs, F. Schuchardt, A. Hennemuth, A. Harloff, Measurement of cardiac valve and aortic blood flow velocities in stroke patients: a comparison of 4d flow mri and echocardiography, The International Journal of Cardiovascular Imaging 34 (6) (2018) 939–946.
166. T. Sugiura, Y. Matsumura, H. Takeuchi, T. Okumiya, Intravascular hemolysis in patients with aortic stenosis: evaluation by erythrocyte creatine, Journal of the American College of Cardiology 65 (10 Supplement) (2015) A1975.
167. P. Garg, R.J. van der Geest, P.P. Swoboda, S. Crandon, G.J. Fent, J.R. Foley, L.E. Dobson, T. Al Musa, S. Onciul, S. Vijayan, et al., Left ventricular thrombus formation in myocardial infarction is associated with altered left ventricular blood flow energetics, European Heart Journal-Cardiovascular Imaging 20 (1) (2018) 108–117.
168. M. Hofer, G. Rappitsch, K. Perktold, W. Trubel, H. Schima, Numerical study of wall mechanics and fluid dynamics in end-to-side anastomoses and correlation to intimal hyperplasia, Journal of Biomechanics 29 (10) (1996) 1297–1308.
169. J.-F. Gerbeau, M. Vidrascu, P. Frey, Fluid–structure interaction in blood flows on geometries based on medical imaging, Computers & Structures 83 (2–3) (2005) 155–165.
170. B. Wolters, M. Rutten, G. Schurink, U. Kose, J. De Hart, F. Van De Vosse, A patient-specific computational model of fluid–structure interaction in abdominal aortic aneurysms, Medical Engineering & Physics 27 (10) (2005) 871–883.
171. R. Torii, M. Oshima, T. Kobayashi, K. Takagi, T.E. Tezduyar, Fluid–structure interaction modeling of a patient-specific cerebral aneurysm: influence of structural modeling, Computational Mechanics 43 (1) (2008) 151.
172. Y. Bazilevs, V.M. Calo, Y. Zhang, T.J. Hughes, Isogeometric fluid–structure interaction analysis

with applications to arterial blood flow, Computational Mechanics 38 (4–5) (2006) 310–322.

173. C.A. Figueroa, I.E. Vignon-Clementel, K.E. Jansen, T.J. Hughes, C.A. Taylor, A coupled momentum method for modeling blood flow in three-dimensional deformable arteries, Computer Methods in Applied Mechanics and Engineering 195 (41–43) (2006) 5685–5706.
174. P. Crosetto, P. Reymond, S. Deparis, D. Kontaxakis, N. Stergiopulos, A. Quarteroni, Fluid–structure interaction simulation of aortic blood flow, Computers & Fluids 43 (1) (2011) 46–57.
175. Dongbin Xiu, Didier Lucor, C.-H. Su, George Em Karniadakis, Stochastic modeling of flow-structure interactions using generalized polynomial chaos, Journal of Fluids Engineering 124 (1) (2001) 51–59.
176. Viorel Mihalef, Dimitris Metaxas, Mark Sussman, Vassilios Hurmusiadis, Leon Axel, Atrioventricular blood flow simulation based on patient-specific data, in: International Conference on Functional Imaging and Modeling of the Heart, Springer, 2009, pp. 386–395.
177. D.A. Steinman, C.A. Taylor, Flow imaging and computing: large artery hemodynamics, Annals of Biomedical Engineering 33 (12) (2005) 1704–1709.
178. J.M. Sohns, J.T. Kowallick, A.A. Joseph, K.D. Merboldt, D. Voit, M. Fasshauer, W. Staab, J. Frahm, J. Lotz, C. Unterberg-Buchwald, Peak flow velocities in the ascending aorta—real-time phase-contrast magnetic resonance imaging vs cine magnetic resonance imaging and echocardiography, Quantitative Imaging in Medicine and Surgery 5 (5) (2015) 685.
179. J. Ryval, A. Straatman, D. Steinman, Two-equation turbulence modeling of pulsatile flow in a stenosed tube, Journal of Biomechanical Engineering 126 (5) (2004) 625–635.
180. F. Hellmeier, S. Nordmeyer, P. Yevtushenko, J. Bruening, F. Berger, T. Kuehne, L. Goubergrits, M. Kelm, Hemodynamic evaluation of a biological and mechanical aortic valve prosthesis using patient-specific mri-based cfd, Artificial Organs 42 (1) (2018) 49–57.
181. L. Ge, H.-L. Leo, F. Sotiropoulos, A.P. Yoganathan, Flow in a mechanical bileaflet heart valve at laminar and near-peak systole flow rates: Cfd simulations and experiments, Journal of Biomechanical Engineering 127 (5) (2005) 782–797.
182. C. Karmonik, S. Partovi, M. Loebe, B. Schmack, A. Weymann, A.B. Lumsden, M. Karck, A. Ruhparwar, Computational fluid dynamics in patients with continuous-flow left ventricular assist device support show hemodynamic alterations in the ascending aorta, Journal of Thoracic and Cardiovascular Surgery 147 (4) (2014) 1326–1333.
183. R. Mittal, J.H. Seo, V. Vedula, Y.J. Choi, H. Liu, H.H. Huang, S. Jain, L. Younes, T. Abraham, R.T. George, Computational modeling of cardiac hemodynamics: current status and future outlook, Journal of Computational Physics 305 (2016) 1065–1082.
184. P.J. Kilner, G.-Z. Yang, A.J. Wilkes, R.H. Mohiaddin, D.N. Firmin, M.H. Yacoub, Asymmetric redirection of flow through the heart, Nature 404 (6779) (2000) 759.
185. N. Stergiopulos, J.-J. Meister, N. Westerhof, Simple and accurate way for estimating total and segmental arterial compliance: the pulse pressure method, Annals of Biomedical Engineering 22 (4) (1994) 392–397.
186. P. Segers, E. Rietzschel, M. De Buyzere, N. Stergiopulos, N. Westerhof, L. Van Bortel, T. Gillebert, P. Verdonck, Three-and four-element windkessel models: assessment of their fitting performance in a large cohort of healthy middle-aged individuals, Proceedings of the Institution of Mechanical Engineers. Part H, Journal of Engineering in Medicine 222 (4) (2008) 417–428.
187. M.J. Powell, The BOBYQA Algorithm for Bound Constrained Optimization Without Derivatives, Cambridge NA Report NA2009/06, 2009.
188. K.C. Wong, M. Sermesant, K. Rhode, M. Ginks, C.A. Rinaldi, R. Razavi, H. Delingette, N. Ayache, Velocity-based cardiac contractility personalization from images using derivative-free optimization, Journal of the Mechanical Behavior of Biomedical Materials 43 (2015) 35–52.
189. L. Le Folgoc, H. Delingette, A. Criminisi, N. Ayache, Current-based 4d shape analysis for the mechanical personalization of heart models, in: International MICCAI Workshop on Medical Computer Vision, Springer, 2012, pp. 283–292.
190. H. Gao, W. Li, L. Cai, C. Berry, X. Luo, Parameter estimation in a Holzapfel–Ogden law for healthy myocardium, Journal of Engineering Mathematics 95 (1) (2015) 231–248.
191. M. Hadjicharalambous, L. Asner, R. Chabiniok, E. Sammut, J. Wong, D. Peressutti, E. Kerfoot, A. King, J. Lee, R. Razavi, et al., Non-invasive model-based assessment of passive left-ventricular myocardial stiffness in healthy subjects and in patients with non-ischemic dilated cardiomyopathy, Annals of Biomedical Engineering 45 (3) (2017) 605–618.
192. J. Xi, P. Lamata, S. Niederer, S. Land, W. Shi, X. Zhuang, S. Ourselin, S.G. Duckett, A.K. Shetty, C.A.

Rinaldi, et al., The estimation of patient-specific cardiac diastolic functions from clinical measurements, Medical Image Analysis 17 (2) (2013) 133–146.

193. P. Chinchapatnam, K.S. Rhode, M. Ginks, C.A. Rinaldi, P. Lambiase, R. Razavi, S. Arridge, M. Sermesant, Model-based imaging of cardiac apparent conductivity and local conduction velocity for diagnosis and planning of therapy, IEEE Transactions on Medical Imaging 27 (11) (2008) 1631–1642.

194. J. Relan, P. Chinchapatnam, M. Sermesant, K. Rhode, M. Ginks, H. Delingette, C.A. Rinaldi, R. Razavi, N. Ayache, Coupled personalization of cardiac electrophysiology models for prediction of ischaemic ventricular tachycardia, Interface Focus 1 (3) (2011) 396–407.

195. P. Seegerer, T. Mansi, M.-P. Jolly, D. Neumann, B. Georgescu, A. Kamen, E. Kayvanpour, A. Amr, F. Sedaghat-Hamedani, J. Haas, H. Katus, B. Meder, D. Comaniciu, Estimation of regional electrical properties of the heart from 12-lead ECG and images, in: Statistical Atlases and Computational Models of the Heart – Imaging and Modelling Challenges, in: LNCS, vol. 8896, Springer, 2015, pp. 204–212.

196. H. Delingette, F. Billet, K.C. Wong, M. Sermesant, K. Rhode, M. Ginks, C.A. Rinaldi, R. Razavi, N. Ayache, Personalization of cardiac motion and contractility from images using variational data assimilation, IEEE Transactions on Biomedical Engineering 59 (1) (2011) 20–24.

197. P. Moireau, D. Chapelle, P. Le Tallec, Joint state and parameter estimation for distributed mechanical systems, Computer Methods in Applied Mechanics and Engineering 197 (6–8) (2008) 659–677.

198. R. Chabiniok, P. Moireau, P.-F. Lesault, A. Rahmouni, J.-F. Deux, D. Chapelle, Estimation of tissue contractility from cardiac cine-mri using a biomechanical heart model, Biomechanics and Modeling in Mechanobiology 11 (5) (2012) 609–630.

199. A. Prakosa, M. Sermesant, P. Allain, N. Villain, C.A. Rinaldi, K. Rhode, R. Razavi, H. Delingette, N. Ayache, Cardiac electrophysiological activation pattern estimation from images using a patient-specific database of synthetic image sequences, IEEE Transactions on Biomedical Engineering 61 (2) (2013) 235–245.

200. O. Camara, A. Pashaei, R. Sebastian, A.F. Frangi, Personalization of fast conduction Purkinje system in Eikonal-based electrophysiological models with optical mapping data, in: International Workshop on Statistical Atlases and Computational Models of the Heart, Springer, 2010, pp. 281–290.

201. E. Konukoglu, J. Relan, U. Cilingir, B.H. Menze, P. Chinchapatnam, A. Jadidi, H. Cochet, M. Hocini, H. Delingette, P. Jaïs, M. Haïssaguerre, N. Ayache, M. Sermesant, Efficient probabilistic model personalization integrating uncertainty on data and parameters: application to Eikonal-diffusion models in cardiac electrophysiology, Progress in Biophysics and Molecular Biology 107 (1) (2011) 134–146.

202. B. Georgescu, X.S. Zhou, D. Comaniciu, A. Gupta, Database-Guided Segmentation of Anatomical Structures With Complex Appearance, 2005 IEEE Computer Society Conference on Computer Vision and Pattern Recognition (CVPR'05), vol. 2, IEEE, 2005, pp. 429–436.

203. Z. Tu, Probabilistic boosting-tree: learning discriminative models for classification, recognition, and clustering, in: Proceedings of the IEEE International Conference on Computer Vision, IEEE Computer Society, 2005, pp. 1589–1596.

204. R.H. Anderson, Clinical anatomy of the aortic root, Heart 84 (6) (2000) 670–673.

205. C. Stamm, R.H. Anderson, S.Y. Ho, Clinical anatomy of the normal pulmonary root compared with that in isolated pulmonary valvular stenosis, Journal of the American College of Cardiology 31 (6) (1998) 1420–1425.

206. A.H.A.W.G. on Myocardial Segmentation, R. for Cardiac Imaging, M.D. Cerqueira, N.J. Weissman, V. Dilsizian, A.K. Jacobs, S. Kaul, W.K. Laskey, D.J. Pennell, J.A. Rumberger, T. Ryan, et al., Standardized myocardial segmentation and nomenclature for tomographic imaging of the heart: a statement for healthcare professionals from the cardiac imaging committee of the council on clinical cardiology of the American heart association, Circulation 105 (4) (2002) 539–542.

207. V. Arsigny, P. Fillard, X. Pennec, N. Ayache, Log-Euclidean metrics for fast and simple calculus on diffusion tensors, Magnetic Resonance in Medicine: An Official Journal of the International Society for Magnetic Resonance in Medicine 56 (2) (2006) 411–421.

208. A.J. Pullan, L.K. Cheng, M.L. Buist, Mathematically Modeling the Electrical Activity of the Heart: From Cell to Body Surface and Back, World Scientific Publishing Company, 2005.

209. R. Clayton, O. Bernus, E. Cherry, H. Dierckx, F. Fenton, L. Mirabella, A. Panfilov, F. Sachse, G. Seemann, H. Zhang, Models of cardiac tissue electrophysiology: progress, challenges and open

questions, Progress in Biophysics and Molecular Biology 104 (1) (2011) 22–48.
210. J. Pitt-Francis, P. Pathmanathan, M.O. Bernabeu, R. Bordas, J. Cooper, A.G. Fletcher, G.R. Mirams, P. Murray, J.M. Osborne, A. Walter, et al., Chaste: a test-driven approach to software development for biological modelling, Computer Physics Communications 180 (12) (2009) 2452–2471.
211. C. Bradley, A. Bowery, R. Britten, V. Budelmann, O. Camara, R. Christie, A. Cookson, A.F. Frangi, T.B. Gamage, T. Heidlauf, et al., Opencmiss: a multi-physics & multi-scale computational infrastructure for the vph/physiome project, Progress in Biophysics and Molecular Biology 107 (1) (2011) 32–47.
212. H. Talbot, S. Marchesseau, C. Duriez, M. Sermesant, S. Cotin, H. Delingette, Towards an interactive electromechanical model of the heart, Interface Focus 3 (2) (2013).
213. P. Pathmanathan, M. Bernabeu, S. Niederer, D. Gavaghan, D. Kay, Computational modelling of cardiac electrophysiology: explanation of the variability of results from different numerical solvers, International Journal for Numerical Methods in Biomedical Engineering 28 (8) (2012) 890–903.
214. S. Chen, G.D. Doolen, Lattice Boltzmann method for fluid flows, Annual Review of Fluid Mechanics 30 (1) (1998) 329–364.
215. C.K. Aidun, J.R. Clausen, Lattice-Boltzmann method for complex flows, Annual Review of Fluid Mechanics 42 (1) (2010) 439–472.
216. H. Yoshida, M. Nagaoka, Multiple-relaxation-time lattice Boltzmann model for the convection and anisotropic diffusion equation, Journal of Computational Physics 229 (October 2010) 7774–7795.
217. S. Rapaka, T. Mansi, B. Georgescu, M. Pop, G.A. Wright, A. Kamen, D. Comaniciu, Lbm-ep: lattice-Boltzmann method for fast cardiac electrophysiology simulation from 3d images, in: Medical Image Computing and Computer-Assisted Intervention–MICCAI 2012, Springer LNCS, 2012, pp. 33–40.
218. B. Georgescu, S. Rapaka, T. Mansi, O. Zettinig, A. Kamen, D. Comaniciu, Towards real-time cardiac electrophysiology computations using gp-gpu lattice-Boltzmann method, in: High-Performance Computing, a 2013 MICCAI Workshop, 2013.
219. J. Campos, R. Oliveira, R. dos Santos, B. Rocha, Lattice Boltzmann method for parallel simulations of cardiac electrophysiology using gpus, Journal of Computational and Applied Mathematics (2015).
220. O. Zettinig, T. Mansi, D. Neumann, B. Georgescu, S. Rapaka, P. Seegerer, E. Kayvanpour, F. Sedaghat-Hamedani, A. Amr, J. Haas, et al., Data-driven estimation of cardiac electrical diffusivity from 12-lead ecg signals, Medical Image Analysis 18 (8) (2014) 1361–1376.
221. S.A. Niederer, E. Kerfoot, A.P. Benson, M.O. Bernabeu, O. Bernus, C. Bradley, E.M. Cherry, R. Clayton, F.H. Fenton, A. Garny, et al., Verification of cardiac tissue electrophysiology simulators using an n-version benchmark, Philosophical Transactions of the Royal Society A: Mathematical, Physical and Engineering Sciences 369 (1954) (2011) 4331–4351.
222. R.C. Barr, M. Ramsey, M.S. Spach, Relating epicardial to body surface potential distributions by means of transfer coefficients based on geometry measurements, IEEE Transactions on Biomedical Engineering 1 (1977) 1–11.
223. B.-U. Kohler, C. Hennig, R. Orglmeister, The principles of software qrs detection, IEEE Engineering in Medicine and Biology Magazine 21 (1) (2002) 42–57.
224. K.C. Wong, L. Wang, H. Zhang, H. Liu, P. Shi, Meshfree implementation of individualized active cardiac dynamics, Computerized Medical Imaging and Graphics 34 (1) (2010) 91–103.
225. È. Lluch, M. De Craene, B. Bijnens, M. Sermesant, J. Noailly, O. Camara, H.G. Morales, Breaking the state of the heart: meshless model for cardiac mechanics, Biomechanics and Modeling in Mechanobiology (2019) 1–13.
226. Z.A. Taylor, M. Cheng, S. Ourselin, High-speed nonlinear finite element analysis for surgical simulation using graphics processing units, IEEE Transactions on Medical Imaging 27 (5) (2008) 650–663.
227. A.J. Chorin, On the convergence of discrete approximations to the Navier–Stokes equations, Mathematics of Computation 23 (106) (1969) 341–353.
228. C. Greenshields, H. Weller, A unified formulation for continuum mechanics applied to fluid–structure interaction in flexible tubes, International Journal for Numerical Methods in Engineering 64 (12) (2005) 1575–1593.
229. C.A. Taylor, T.A. Fonte, J.K. Min, Computational fluid dynamics applied to cardiac computed tomography for noninvasive quantification of fractional flow reserve: scientific basis, Journal of the American College of Cardiology 61 (22) (2013) 2233–2241.
230. E. Kung, A. Baretta, C. Baker, G. Arbia, G. Biglino, C. Corsini, S. Schievano, I.E. Vignon-Clementel, G.

Dubini, G. Pennati, et al., Predictive modeling of the virtual hemi-fontan operation for second stage single ventricle palliation: two patient-specific cases, Journal of Biomechanics 46 (2) (2013) 423–429.
231. C. Long, M.-C. Hsu, Y. Bazilevs, J. Feinstein, A. Marsden, Fluid–structure interaction simulations of the fontan procedure using variable wall properties, International Journal for Numerical Methods in Biomedical Engineering 28 (5) (2012) 513–527.
232. D. Sengupta, A.M. Kahn, J.C. Burns, S. Sankaran, S.C. Shadden, A.L. Marsden, Image-based modeling of hemodynamics in coronary artery aneurysms caused by Kawasaki disease, Biomechanics and Modeling in Mechanobiology 11 (6) (2012) 915–932.
233. W.K. Liu, Y. Liu, D. Farrell, L. Zhang, X.S. Wang, Y. Fukui, N. Patankar, Y. Zhang, C. Bajaj, J. Lee, et al., Immersed finite element method and its applications to biological systems, Computer Methods in Applied Mechanics and Engineering 195 (13–16) (2006) 1722–1749.
234. H. Watanabe, S. Sugiura, H. Kafuku, T. Hisada, Multiphysics simulation of left ventricular filling dynamics using fluid-structure interaction finite element method, Biophysical Journal 87 (3) (2004) 2074–2085.
235. A.D. Pasipoularides, M. Shu, M.S. Womack, A. Shah, O. Von Ramm, D.D. Glower, Rv functional imaging: 3-d echo-derived dynamic geometry and flow field simulations, American Journal of Physiology. Heart and Circulatory Physiology 284 (1) (2003) H56–H65.
236. J. Nichols, S. Lele, P. Moin, F. Ham, G. Brès, J. Bridges, Large-eddy simulation for supersonic rectangular jet noise prediction: effects of chevrons, in: 18th AIAA/CEAS Aeroacoustics Conference (33rd AIAA Aeroacoustics Conference), 2012, p. 2212.
237. T. Schenkel, M. Malve, M. Reik, M. Markl, B. Jung, H. Oertel, Mri-based cfd analysis of flow in a human left ventricle: methodology and application to a healthy heart, Annals of Biomedical Engineering 37 (3) (2009) 503–515.
238. Y. Cheng, H. Oertel, T. Schenkel, Fluid-structure coupled cfd simulation of the left ventricular flow during filling phase, Annals of Biomedical Engineering 33 (5) (2005) 567–576.
239. N.R. Saber, N.B. Wood, A. Gosman, R.D. Merrifield, G.-Z. Yang, C.L. Charrier, P.D. Gatehouse, D.N. Firmin, Progress towards patient-specific computational flow modeling of the left heart via combination of magnetic resonance imaging with computational fluid dynamics, Annals of Biomedical Engineering 31 (1) (2003) 42–52.
240. J.H. Seo, V. Vedula, T. Abraham, A.C. Lardo, F. Dawoud, H. Luo, R. Mittal, Effect of the mitral valve on diastolic flow patterns, Physics of Fluids 26 (12) (2014) 121901.
241. S. Hendabadi, J. Bermejo, Y. Benito, R. Yotti, F. Fernández-Avilés, J.C. Del Álamo, S.C. Shadden, Topology of blood transport in the human left ventricle by novel processing of Doppler echocardiography, Annals of Biomedical Engineering 41 (12) (2013) 2603–2616.
242. R. Delewi, F. Zijlstra, J.J. Piek, Left ventricular thrombus formation after acute myocardial infarction, Heart 98 (23) (2012) 1743–1749.
243. B.E. Griffith, X. Luo, D.M. McQueen, C.S. Peskin, Simulating the fluid dynamics of natural and prosthetic heart valves using the immersed boundary method, International Journal of Applied Mechanics 1 (01) (2009) 137–177.
244. V. Babaliaros, A. Cribier, C. Agatiello, Surgery insight: current advances in percutaneous heart valve replacement and repair, Nature Reviews Cardiology 3 (5) (2006) 256.
245. M. Courtois, S.J. Kovács Jr, P. Ludbrook, Transmitral pressure-flow velocity relation. importance of regional pressure gradients in the left ventricle during diastole, Circulation 78 (3) (1988) 661–671.
246. Y. Ishida, J. Meisner, K. Tsujioka, J. Gallo, C. Yoran, R. Frater, E. Yellin, Left ventricular filling dynamics: influence of left ventricular relaxation and left atrial pressure, Circulation 74 (1) (1986) 187–196.
247. B.M. Yun, L. Dasi, C. Aidun, A. Yoganathan, Computational modelling of flow through prosthetic heart valves using the entropic lattice-Boltzmann method, Journal of Fluid Mechanics 743 (2014) 170–201.
248. R. Mittal, G. Iaccarino, Immersed boundary methods, Annual Review of Fluid Mechanics 37 (2005) 239–261.
249. C.S. Peskin, The immersed boundary method, Acta Numerica 11 (2002) 479–517.
250. C.S. Peskin, Numerical analysis of blood flow in the heart, Journal of Computational Physics 25 (3) (1977) 220–252.
251. Y. Zang, R.L. Street, J.R. Koseff, A non-staggered grid, fractional step method for time-dependent incompressible Navier–Stokes equations in curvilinear coordinates, Journal of Computational Physics 114 (1) (1994) 18–33.
252. Dominique d'Humieres, Multiple-relaxation-time lattice Boltzmann models in three dimensions,

Philosophical Transactions of the Royal Society of London. Series A: Mathematical, Physical and Engineering Sciences 360 (1792) (2002) 437–451.
253. Santosh Ansumali, Iliya V. Karlin, Hans Christian Öttinger, Minimal entropic kinetic models for hydrodynamics, EPL (Europhysics Letters) 63 (6) (2003) 798.
254. M. Bouzidi, M. Firdaouss, P. Lallemand, Momentum transfer of a Boltzmann-lattice fluid with boundaries, Physics of Fluids 13 (11) (2001) 3452–3459.
255. Guo Zhao Li, Chu-Guang Zheng, Bao-Chang Shi, Non-equilibrium extrapolation method for velocity and pressure boundary conditions in the lattice Boltzmann method, Chinese Physics 11 (4) (2002) 366.
256. Kevin Connington, Qinjun Kang, Hari Viswanathan, Amr Abdel-Fattah, Shiyi Chen, Peristaltic particle transport using the lattice Boltzmann method, Physics of Fluids 21 (5) (2009) 053301.
257. F.C. Ghesu, E. Krubasik, B. Georgescu, V. Singh, Y. Zheng, J. Hornegger, D. Comaniciu, Marginal space deep learning: efficient architecture for volumetric image parsing, IEEE Transactions on Medical Imaging 35 (May 2016) 1217–1228.
258. F.C. Ghesu, B. Georgescu, Y. Zheng, J. Hornegger, D. Comaniciu, Marginal space deep learning: efficient architecture for detection in volumetric image data, in: Medical Image Computing and Computer-Assisted Intervention, in: LNCS, vol. 9349, Springer International Publishing, Cham, Switzerland, Oct. 2015, pp. 710–718.
259. P. Viola, M. Jones, Rapid object detection using a boosted cascade of simple features, in: Proceedings of the IEEE Conference on Computer Vision and Pattern Recognition, IEEE Computer Society, 2001, pp. 511–518.
260. F.C. Ghesu, B. Georgescu, J. Hornegger, Deep Learning for Medical Image Analysis, Ch. Efficient medical image parsing, Academic Press, New York, USA, Jan. 2017, pp. 55–81.
261. V. Mnih, K. Kavukcuoglu, D. Silver, A.A. Rusu, J. Veness, M.G. Bellemare, A. Graves, M. Riedmiller, A.K. Fidjeland, G. Ostrovski, S. Petersen, C. Beattie, A. Sadik, I. Antonoglou, H. King, D. Kumara, D. Wierstra, S. Legg, D. Hassabis, Human-level control through deep reinforcement learning, Nature 518 (7540) (2015) 529–533.
262. F.C. Ghesu, B. Georgescu, Y. Zheng, S. Grbic, A. Maier, J. Hornegger, D. Comaniciu, Multi-scale deep reinforcement learning for real-time 3D-landmark detection in CT scans, IEEE Transactions on Pattern Analysis and Machine Intelligence 41 (1) (2017) 176–189.
263. F.C. Ghesu, B. Georgescu, S. Grbic, A. Maier, J. Hornegger, D. Comaniciu, Towards intelligent robust detection of anatomical structures in incomplete volumetric data, Medical Image Analysis 48 (Aug. 2018) 203–213.
264. F.C. Ghesu, B. Georgescu, S. Grbic, A.K. Maier, J. Hornegger, D. Comaniciu, Robust multi-scale anatomical landmark detection in incomplete 3D-CT data, in: Medical Image Computing and Computer Assisted Intervention, in: LNCS, vol. 10433, Springer International Publishing, Cham, Switzerland, Sept. 2017, pp. 194–202.
265. F.C. Ghesu, B. Georgescu, T. Mansi, D. Neumann, J. Hornegger, D. Comaniciu, An artificial agent for anatomical landmark detection in medical images, in: International Conference on Medical Image Computing and Computer-Assisted Intervention, Springer, 2016, pp. 229–237.
266. R.S. Sutton, A.G. Barto, Reinforcement Learning: An Introduction, vol. 1, MIT Press, Cambridge, 1998.
267. C.J.C.H. Watkins, P. Dayan, Q-learning, Machine Learning 8 (May 1992) 279–292.
268. C.J.C.H. Watkins, Learning From Delayed Rewards, PhD thesis, King's College, Cambridge, England, May 1989.
269. T. Lindeberg, Scale-Space Theory in Computer Vision, Kluwer Academic Publishers, Norwell, MA, USA, 1994.
270. R. Bellman, Dynamic Programming, Princeton University Press, 1957.
271. P.H.S. Torr, A. Zisserman, MLESAC: a new robust estimator with application to estimating image geometry, Computer Vision and Image Understanding 78 (Apr. 2000) 138–156.
272. X. Lu, B. Georgescu, M.-P. Jolly, J. Guehring, A. Young, B. Cowan, A. Littmann, D. Comaniciu, Cardiac anchoring in MRI through context modeling, in: T. Jiang, N. Navab, J.P.W. Pluim, M.A. Viergever (Eds.), MICCAI 2010, Part I, in: LNCS, vol. 6361, Springer, Heidelberg, 2010, pp. 383–390.
273. X. Lu, M.-P. Jolly, Discriminative context modeling using auxiliary markers for LV landmark detection from a single MR image, in: O. Camara, T. Mansi, M. Pop, K. Rhode, M. Sermesant, A. Young (Eds.), TACOM 2013, in: LNCS, vol. 7746, Springer, Heidelberg, 2013, pp. 105–114.
274. J. Long, E. Shelhamer, T. Darrell, Fully convolutional networks for semantic segmentation, in: IEEE CVPR, 2015, pp. 3431–3440.
275. Ö. Çiçek, A. Abdulkadir, S.S. Lienkamp, T. Brox, O. Ronneberger, 3d U-Net: learning dense volumetric segmentation from sparse annotation, in: MICCAI, Springer, 2016, pp. 424–432.

276. G. Huang, Z. Liu, L. Van Der Maaten, K.Q. Weinberger, Densely connected convolutional networks, in: IEEE CVPR, 2017, pp. 4700–4708.
277. J.D. Dormer, L. Ma, M. Halicek, C.M. Reilly, E. Schreibmann, B. Fei, Heart chamber segmentation from ct using convolutional neural networks, in: Medical Imaging: Biomedical Applications in Molecular, Structural, and Functional Imaging, vol. 10578, 2018.
278. Q. Zheng, H. Delingette, N. Duchateau, N. Ayache, 3-D consistent and robust segmentation of cardiac images by deep learning with spatial propagation, IEEE Transactions on Medical Imaging 37 (9) (2018) 2137–2148.
279. C. Bercea, O. Pauly, A.K. Maier, F.C. Ghesu, SHAMANN: shared memory augmented neural networks, in: IPMI, 2019, in press.
280. D. Yang, D. Xu, S.K. Zhou, B. Georgescu, M. Chen, S. Grbic, D. Metaxas, D. Comaniciu, Automatic liver segmentation using an adversarial image-to-image network, in: MICCAI, Springer, 2017, pp. 507–515.
281. Y. Wang, B. Georgescu, T. Chen, W. Wu, P. Wang, X. Lu, R. Ionasec, Y. Zheng, D. Comaniciu, Learning-based detection and tracking in medical imaging: a probabilistic approach, in: Deformation Models, Springer, 2013, pp. 209–235.
282. A. Yilmaz, O. Javed, M. Shah, Object tracking: a survey, ACM Computing Surveys (CSUR) 38 (4) (2006) 13.
283. Y. Wu, J. Lim, M.-H. Yang, Online object tracking: a benchmark, in: Proceedings of the IEEE Conference on Computer Vision and Pattern Recognition, 2013, pp. 2411–2418.
284. H. Yang, L. Shao, F. Zheng, L. Wang, Z. Song, Recent advances and trends in visual tracking: a review, Neurocomputing 74 (18) (2011) 3823–3831.
285. M. Fiaz, A. Mahmood, S.K. Jung, Tracking noisy targets: a review of recent object tracking approaches, arXiv preprint, arXiv:1802.03098, 2018.
286. J. Bromley, I. Guyon, Y. LeCun, E. Säckinger, R. Shah, Signature verification using a "Siamese" time delay neural network, in: Advances in Neural Information Processing Systems, 1994, pp. 737–744.
287. L. Bertinetto, J. Valmadre, J.F. Henriques, A. Vedaldi, P.H. Torr, Fully-convolutional Siamese networks for object tracking, in: European Conference on Computer Vision, Springer, 2016, pp. 850–865.
288. D. Held, S. Thrun, S. Savarese, Learning to track at 100 fps with deep regression networks, in: European Conference Computer Vision (ECCV), 2016.
289. Q. Guo, W. Feng, C. Zhou, R. Huang, L. Wan, S. Wang, Learning dynamic Siamese network for visual object tracking, in: 2017 IEEE International Conference on Computer Vision (ICCV), IEEE, 2017, pp. 1781–1789.
290. J. Valmadre, L. Bertinetto, J. Henriques, A. Vedaldi, P.H. Torr, End-to-end representation learning for correlation filter based tracking, in: Computer Vision and Pattern Recognition (CVPR), 2017 IEEE Conference on, IEEE, 2017, pp. 5000–5008.
291. Q. Wang, J. Gao, J. Xing, M. Zhang, W. Hu, Dcfnet: discriminant correlation filters network for visual tracking, arXiv preprint, arXiv:1704.04057, 2017.
292. B. Li, J. Yan, W. Wu, Z. Zhu, X. Hu, High performance visual tracking with Siamese region proposal network, in: Proceedings of the IEEE Conference on Computer Vision and Pattern Recognition, 2018, pp. 8971–8980.
293. B. Li, W. Wu, Q. Wang, F. Zhang, J. Xing, J. Yan, Siamrpn++: evolution of Siamese visual tracking with very deep networks, arXiv preprint, arXiv:1812.11703, 2018.
294. Z. Zhu, Q. Wang, B. Li, W. Wu, J. Yan, W. Hu, Distractor-aware Siamese networks for visual object tracking, in: Proceedings of the European Conference on Computer Vision (ECCV), 2018, pp. 101–117.
295. H. Nam, B. Han, Learning multi-domain convolutional neural networks for visual tracking, in: The IEEE Conference on Computer Vision and Pattern Recognition (CVPR), June 2016.
296. Y. Song, C. Ma, L. Gong, J. Zhang, R.W. Lau, M.-H. Yang, Crest: convolutional residual learning for visual tracking, in: Computer Vision (ICCV), 2017 IEEE International Conference on, IEEE, 2017, pp. 2574–2583.
297. D.S. Bolme, J.R. Beveridge, B.A. Draper, Y.M. Lui, Visual object tracking using adaptive correlation filters, in: Computer Vision and Pattern Recognition (CVPR), 2010 IEEE Conference on, IEEE, 2010, pp. 2544–2550.
298. M. Danelljan, G. Hager, F. Shahbaz Khan, M. Felsberg, Convolutional features for correlation filter based visual tracking, in: Proceedings of the IEEE International Conference on Computer Vision Workshops, 2015, pp. 58–66.
299. Z. Zhu, W. Wu, W. Zou, J. Yan, End-to-end flow correlation tracking with spatial-temporal attention, Illumination 42 (2018) 20.
300. N. Parajuli, A. Lu, J.C. Stendahl, M. Zontak, N. Boutagy, I. Alkhalil, M. Eberle, B.A. Lin, M. O'Donnell, A.J. Sinusas, et al., Flow network based cardiac motion tracking leveraging learned feature matching, in: International Conference on Medical Image Computing and

Computer-Assisted Intervention, Springer, 2017, pp. 279–286.
301. Z. Cui, S. Xiao, J. Feng, S. Yan, Recurrently target-attending tracking, in: Proceedings of the IEEE Conference on Computer Vision and Pattern Recognition, 2016, pp. 1449–1458.
302. P. Ondruska, I. Posner, Deep tracking: seeing beyond seeing using recurrent neural networks, arXiv preprint, arXiv:1602.00991, 2016.
303. S.E. Kahou, V. Michalski, R. Memisevic, Ratm: recurrent attentive tracking model, in: Proc. IEEE Conf. Comput. Vis. Pattern Recognit. Workshops, 2015, pp. 1613–1622.
304. Q. Gan, Q. Guo, Z. Zhang, K. Cho, First step toward model-free, anonymous object tracking with recurrent neural networks, arXiv preprint, arXiv:1511.06425, 2015.
305. G. Ning, Z. Zhang, C. Huang, X. Ren, H. Wang, C. Cai, Z. He, Spatially supervised recurrent convolutional neural networks for visual object tracking, in: Circuits and Systems (ISCAS), 2017 IEEE International Symposium on, IEEE, 2017, pp. 1–4.
306. D. Zhang, H. Maei, X. Wang, Y.-F. Wang, Deep reinforcement learning for visual object tracking in videos, arXiv preprint, arXiv:1701.08936, 2017.
307. J. Choi, J. Kwon, K.M. Lee, Visual tracking by reinforced decision making, arXiv preprint, arXiv:1702.06291, 2017.
308. C. Huang, S. Lucey, D. Ramanan, Learning policies for adaptive tracking with deep feature cascades, in: IEEE Int. Conf. on Computer Vision (ICCV), 2017, pp. 105–114.
309. S. Yun, J. Choi, Y. Yoo, K. Yun, J. Young Choi, Action-decision networks for visual tracking with deep reinforcement learning, in: Proceedings of the IEEE Conference on Computer Vision and Pattern Recognition, 2017, pp. 2711–2720.
310. L. Bertinetto, J. Valmadre, S. Golodetz, O. Miksik, P.H. Torr, Staple: complementary learners for real-time tracking, in: Proceedings of the IEEE Conference on Computer Vision and Pattern Recognition, 2016, pp. 1401–1409.
311. J.F. Henriques, R. Caseiro, P. Martins, J. Batista, High-speed tracking with kernelized correlation filters, IEEE Transactions on Pattern Analysis and Machine Intelligence 37 (3) (2015) 583–596.
312. H.K. Galoogahi, A. Fagg, S. Lucey, Learning background-aware correlation filters for visual tracking, in: ICCV, vol. 3, 2017, p. 4.
313. Y. Song, C. Ma, X. Wu, L. Gong, L. Bao, W. Zuo, C. Shen, R.W. Lau, M.-H. Yang, Vital: visual tracking via adversarial learning, in: Proceedings of the IEEE Conference on Computer Vision and Pattern Recognition, 2018, pp. 8990–8999.
314. V.G. Ng, A.J. Lansky, Novel qca methodologies and angiographic scores, The International Journal of Cardiovascular Imaging 27 (2) (2011) 157–165.
315. L. Itu, P. Sharma, V. Mihalef, A. Kamen, C. Suciu, D. Lomaniciu, A patient-specific reduced-order model for coronary circulation, in: Biomedical Imaging (ISBI), 2012 9th IEEE International Symposium on, IEEE, 2012, pp. 832–835.
316. W.B. Meijboom, C.A. Van Mieghem, N. van Pelt, A. Weustink, F. Pugliese, N.R. Mollet, E. Boersma, E. Regar, R.J. van Geuns, P.J. de Jaegere, et al., Comprehensive assessment of coronary artery stenoses: computed tomography coronary angiography versus conventional coronary angiography and correlation with fractional flow reserve in patients with stable angina, Journal of the American College of Cardiology 52 (8) (2008) 636–643.
317. N.H. Pijls, Fractional flow reserve after previous myocardial infarction, European Heart Journal 28 (19) (2007) 2301–2302.
318. P.A. Tonino, B. De Bruyne, N.H. Pijls, U. Siebert, F. Ikeno, M. vant Veer, V. Klauss, G. Manoharan, T. Engstrøm, K.G. Oldroyd, et al., Fractional flow reserve versus angiography for guiding percutaneous coronary intervention, The New England Journal of Medicine 360 (3) (2009) 213–224.
319. S.D. Fihn, J.M. Gardin, J. Abrams, K. Berra, J.C. Blankenship, P.S. Douglas, J.M. Foody, T.C. Gerber, A.L. Hinderliter, S.B. King, et al., 2012 accf/aha/acp/aats/pcna/scai/sts guideline for the diagnosis and management of patients with stable ischemic heart disease: a report of the American college of cardiology foundation/American heart association task force on practice guidelines, and the American college of physicians, American association for thoracic surgery, preventive cardiovascular nurses association, society for cardiovascular angiography and interventions, and society of thoracic surgeons, Journal of the American College of Cardiology 60 (24) (2012) e44–e164.
320. A.F. members, S. Windecker, P. Kolh, F. Alfonso, J.-P. Collet, J. Cremer, V. Falk, G. Filippatos, C. Hamm, S.J. Head, et al., 2014 esc/eacts guidelines on myocardial revascularization: the task force on myocardial revascularization of the European society of cardiology (esc) and the European association for cardio-thoracic surgery (eacts) developed with the special contribution of the European association of percutaneous cardiovascular interventions (eapci), European Heart Journal 35 (37) (2014) 2541–2619.

321. S. Tu, C.V. Bourantas, B.L. Nørgaard, G.S. Kassab, B.-K. Koo, J. Reiber, Image-based assessment of fractional flow reserve, EuroIntervention: Journal of EuroPCR in Collaboration with the Working Group on Interventional Cardiology of the European Society of Cardiology 11 (2015), V50–4.
322. A. Coenen, M.M. Lubbers, A. Kurata, A. Kono, A. Dedic, R.G. Chelu, M.L. Dijkshoorn, F.J. Gijsen, M. Ouhlous, R.-J.M. van Geuns, et al., Fractional flow reserve computed from noninvasive ct angiography data: diagnostic performance of an on-site clinician-operated computational fluid dynamics algorithm, Radiology 274 (3) (2014) 674–683.
323. B.-K. Koo, H.-M. Yang, J.-H. Doh, H. Choe, S.-Y. Lee, C.-H. Yoon, Y.-K. Cho, C.-W. Nam, S.-H. Hur, H.-S. Lim, et al., Optimal intravascular ultrasound criteria and their accuracy for defining the functional significance of intermediate coronary stenoses of different locations, JACC: Cardiovascular Interventions 4 (7) (2011) 803–811.
324. J.K. Min, J. Leipsic, M.J. Pencina, D.S. Berman, B.-K. Koo, C. van Mieghem, A. Erglis, F.Y. Lin, A.M. Dunning, P. Apruzzese, et al., Diagnostic accuracy of fractional flow reserve from anatomic ct angiography, JAMA 308 (12) (2012) 1237–1245.
325. P.D. Morris, D. Ryan, A.C. Morton, R. Lycett, P.V. Lawford, D.R. Hose, J.P. Gunn, Virtual fractional flow reserve from coronary angiography: modeling the significance of coronary lesions: results from the virtu-1 (virtual fractional flow reserve from coronary angiography) study, JACC: Cardiovascular Interventions 6 (2) (2013) 149–157.
326. B.L. Nørgaard, J. Leipsic, S. Gaur, S. Seneviratne, B.S. Ko, H. Ito, J.M. Jensen, L. Mauri, B. De Bruyne, H. Bezerra, et al., Diagnostic performance of noninvasive fractional flow reserve derived from coronary computed tomography angiography in suspected coronary artery disease: the nxt trial (analysis of coronary blood flow using ct angiography: next steps), Journal of the American College of Cardiology 63 (12) (2014) 1145–1155.
327. M.I. Papafaklis, T. Muramatsu, Y. Ishibashi, L.S. Lakkas, S. Nakatani, C.V. Bourantas, J. Ligthart, Y. Onuma, M. Echavarria-Pinto, G. Tsirka, et al., Fast virtual functional assessment of intermediate coronary lesions using routine angiographic data and blood flow simulation in humans: comparison with pressure wire-fractional flow reserve, EuroIntervention: Journal of EuroPCR in Collaboration with the Working Group on Interventional Cardiology of the European Society of Cardiology 10 (5) (2014) 574–583.
328. M. Renker, U.J. Schoepf, R. Wang, F.G. Meinel, J.D. Rier, R.R. Bayer II, H. Möllmann, C.W. Hamm, D.H. Steinberg, S. Baumann, Comparison of diagnostic value of a novel noninvasive coronary computed tomography angiography method versus standard coronary angiography for assessing fractional flow reserve, American Journal of Cardiology 114 (9) (2014) 1303–1308.
329. S. Tu, E. Barbato, Z. Köszegi, J. Yang, Z. Sun, N.R. Holm, B. Tar, Y. Li, D. Rusinaru, W. Wijns, et al., Fractional flow reserve calculation from 3-dimensional quantitative coronary angiography and timi frame count: a fast computer model to quantify the functional significance of moderately obstructed coronary arteries, JACC: Cardiovascular Interventions 7 (7) (2014) 768–777.
330. S.-B. Deng, X.-D. Jing, J. Wang, C. Huang, S. Xia, J.-L. Du, Y.-J. Liu, Q. She, Diagnostic performance of noninvasive fractional flow reserve derived from coronary computed tomography angiography in coronary artery disease: a systematic review and meta-analysis, International Journal of Cardiology 184 (2015) 703–709.
331. C.M. Bishop, Pattern Recognition and Machine Learning, vol. 4, Springer New York, Berlin, Heidelberg, 2006.
332. M. Zamir, P. Sinclair, T.H. Wonnacott, Relation between diameter and flow in major branches of the arch of the aorta, Journal of Biomechanics 25 (11) (1992) 1303–1310.
333. K. Tøndel, U.G. Indahl, A.B. Gjuvsland, J.O. Vik, P. Hunter, S.W. Omholt, H. Martens, Hierarchical cluster-based partial least squares regression (hc-plsr) is an efficient tool for metamodelling of nonlinear dynamic models, BMC Systems Biology 5 (1) (2011) 90.
334. L. Itu, S. Rapaka, T. Passerini, B. Georgescu, C. Schwemmer, M. Schoebinger, T. Flohr, P. Sharma, D. Comaniciu, A machine-learning approach for computation of fractional flow reserve from coronary computed tomography, Journal of Applied Physiology 121 (1) (2016) 42–52.
335. S. Grosskopf, C. Biermann, K. Deng, Y. Chen, Accurate, fast, and robust vessel contour segmentation of cta using an adaptive self-learning edge model, in: Medical Imaging 2009: Image Processing, vol. 7259, International Society for Optics and Photonics, 2009, p. 72594D.
336. S. El Sayed, E.A. El Sawa, A.E. Atta-Alla, E.A. El, K.H.H. Baassiri, Morphometric study of the right coronary artery, International Journal of Anatomy and Research 3 (3) (2015) 1362–1370.
337. G.S. Kassab, Scaling laws of vascular trees: of form and function, American Journal of Physiology.

Heart and Circulatory Physiology 290 (2) (2006) H894–H903.
338. C.D. Murray, The physiological principle of minimum work: I. the vascular system and the cost of blood volume, Proceedings of the National Academy of Sciences 12 (3) (1926) 207–214.
339. M.S. Olufsen, C.S. Peskin, W.Y. Kim, E.M. Pedersen, A. Nadim, J. Larsen, Numerical simulation and experimental validation of blood flow in arteries with structured-tree outflow conditions, Annals of Biomedical Engineering 28 (11) (2000) 1281–1299.
340. Y. Zhou, G.S. Kassab, S. Molloi, On the design of the coronary arterial tree: a generalization of Murray's law, Physics in Medicine and Biology 44 (12) (1999) 2929.
341. S. Aharinejad, W. Schreiner, F. Neumann, Morphometry of human coronary arterial trees, The Anatomical Record: An Official Publication of the American Association of Anatomists 251 (1) (1998) 50–59.
342. G.S. Kassab, Y.-C.B. Fung, The pattern of coronary arteriolar bifurcations and the uniform shear hypothesis, Annals of Biomedical Engineering 23 (1) (1995) 13–20.
343. R. Wang, M. Renker, U.J. Schoepf, J.L. Wichmann, S.R. Fuller, J.D. Rier, R.R. Bayer II, D.H. Steinberg, C.N. De Cecco, S. Baumann, Diagnostic value of quantitative stenosis predictors with coronary ct angiography compared to invasive fractional flow reserve, European Journal of Radiology 84 (8) (2015) 1509–1515.
344. S. Baumann, R. Wang, U.J. Schoepf, D.H. Steinberg, J.V. Spearman, R.R. Bayer, C.W. Hamm, M. Renker, Coronary ct angiography-derived fractional flow reserve correlated with invasive fractional flow reserve measurements–initial experience with a novel physician-driven algorithm, European Radiology 25 (4) (2015) 1201–1207.
345. A. Coenen, M.M. Lubbers, A. Kurata, A. Kono, A. Dedic, R.G. Chelu, M.L. Dijkshoorn, R.-J.M. van Geuns, M. Schoebinger, L. Itu, et al., Coronary ct angiography derived fractional flow reserve: methodology and evaluation of a point of care algorithm, Journal of Cardiovascular Computed Tomography 10 (2) (2016) 105–113.
346. J. De Geer, M. Sandstedt, A. Björkholm, J. Alfredsson, M. Janzon, J. Engvall, A. Persson, Software-based on-site estimation of fractional flow reserve using standard coronary ct angiography data, Acta Radiologica 57 (10) (2016) 1186–1192.
347. M. Kruk, L. Wardziak, M. Demkow, W. Pleban, J. Pregowski, Z. Dzielinska, M. Witulski, A. Witkowski, W. Ruzyllo, C. Kepka, Workstation-based calculation of cta-based ffr for intermediate stenosis, JACC: Cardiovascular Imaging 9 (6) (2016) 690–699.
348. S. Mantero, R. Pietrabissa, R. Fumero, The coronary bed and its role in the cardiovascular system: a review and an introductory single-branch model, Journal of Biomedical Engineering 14 (2) (1992) 109–116.
349. J.S. Choy, G.S. Kassab, Scaling of myocardial mass to flow and morphometry of coronary arteries, Journal of Applied Physiology 104 (5) (2008) 1281–1286.
350. P. Sharma, L. Itu, X. Zheng, A. Kamen, D. Bernhardt, C. Suciu, D. Comaniciu, A framework for personalization of coronary flow computations during rest and hyperemia, in: Engineering in Medicine and Biology Society (EMBC), 2012 Annual International Conference of the IEEE, IEEE, 2012, pp. 6665–6668.
351. I. Ben-Dor, M. Mahmoudi, T. Deksissa, A.B. Bui, M.A. Gaglia, M.A. Gonzalez, G. Maluenda, G. Sardi, R. Romaguera, A. Laynez-Carnicero, et al., Correlation between fractional flow reserve and intravascular ultrasound lumen area in intermediate coronary artery stenosis, Journal of the American College of Cardiology 57 (14 Supplement) (2011) E1855.
352. N. Gonzalo, J. Escaned, F. Alfonso, C. Nolte, V. Rodriguez, P. Jimenez-Quevedo, C. Bañuelos, A. Fernández-Ortiz, E. Garcia, R. Hernandez-Antolin, et al., Morphometric assessment of coronary stenosis relevance with optical coherence tomography: a comparison with fractional flow reserve and intravascular ultrasound, Journal of the American College of Cardiology 59 (12) (2012) 1080–1089.
353. M.J. Kern, Coronary physiology revisited: practical insights from the cardiac catheterization laboratory, Circulation 101 (11) (2000) 1344–1351.
354. S. Sen, J. Escaned, I.S. Malik, G.W. Mikhail, R.A. Foale, R. Mila, J. Tarkin, R. Petraco, C. Broyd, R. Jabbour, et al., Development and validation of a new adenosine-independent index of stenosis severity from coronary wave-intensity analysis: results of the advise (adenosine vasodilator independent stenosis evaluation) study, Journal of the American College of Cardiology 59 (15) (2012) 1392–1402.
355. T.P. van de Hoef, F. Nolte, P. Damman, R. Delewi, M. Bax, S.A. Chamuleau, M. Voskuil, M. Siebes, J.G. Tijssen, J.A. Spaan, et al., Diagnostic accuracy of combined intracoronary pressure and flow velocity information during baseline conditions:

adenosine-free assessment of functional coronary lesion severity, Circulation: Cardiovascular Interventions (2012), CIRCINTERVENTIONS–111.
356. M. Meuwissen, M. Siebes, S.A. Chamuleau, B.L. van Eck-Smit, K.T. Koch, R.J. de Winter, J.G. Tijssen, J.A. Spaan, J.J. Piek, Hyperemic stenosis resistance index for evaluation of functional coronary lesion severity, Circulation 106 (4) (2002) 441–446.
357. H. Samady, P. Eshtehardi, M.C. McDaniel, J. Suo, S.S. Dhawan, C. Maynard, L.H. Timmins, A.A. Quyyumi, D.P. Giddens, Coronary artery wall shear stress is associated with progression and transformation of atherosclerotic plaque and arterial remodeling in patients with coronary artery disease, Circulation (2011), CIRCULATIONAHA–111.
358. N.H. Pijls, B. de Bruyne, K. Peels, P.H. van der Voort, H.J. Bonnier, J. Bartunek, J.J. Koolen, Measurement of fractional flow reserve to assess the functional severity of coronary-artery stenoses, The New England Journal of Medicine 334 (26) (1996) 1703–1708.
359. M. Courtemanche, R. Ramirez, S. Nattel, Ionic mechanisms underlying human atrial action potential properties: insights from a mathematical model, The American Journal of Physiology 275 (1998) H301–H321.
360. E. Sobie, Parameter sensitivity analysis in electrophysiological models using multivariable regression, Biophysical Journal 96 (4) (2009) 1264–1274.
361. T. Mansi, B. Georgescu, P. Hunter, A. Kamen, D. Comaniciu, Data-driven reduction of a cardiac myofilament model, in: Functional Imaging and Modeling of the Heart, vol. 7945, 2013, pp. 232–240.
362. H. Yang, T. Passerini, T. Mansi, D. Comaniciu, Data-driven model reduction for fast, high fidelity atrial electrophysiology computations, in: Functional Imaging and Modeling of the Heart - 8th International Conference, FIMH 2015 Maastricht, the Netherlands, June 25-27, 2015. Proceedings, 2015, pp. 466–474.
363. F. Atienza, J. Almendral, J. Moreno, R. Vaidyanathan, A. Talkachou, J. Kalifa, A. Arenal, J. Villacastin, E. Torrecilla, A. Sanchez, R. Ploutz-Snyder, J. Jalife, O. Berenfeld, Activation of inward rectifier potassium channels accelerates atrial fibrillation in humans: evidence for a reentrant mechanism, Circulation 114 (2006) 2434–2442.
364. O. Aslanidi, M. Colman, J. Stott, H. Dobrzynski, M. Boyett, A. Holden, H. Zhang, 3d virtual human atria: a computational platform for studying clinical atrial fibrillation, Progress in Biophysics and Molecular Biology 107 (2011) 156–168.
365. J. Friedman, T. Hastie, R. Tibshirani, The Elements of Statistical Learning: Data Mining, Inference, and Prediction, Springer, New York, 2009.
366. S. Roweis, L. Saul, Nonlinear dimensionality reduction by locally linear embedding, Science 290 (5500) (2000) 2323–2326.
367. J. Friedman, W. Stuetzle, Projection pursuit regression, Journal of the American Statistical Association 76 (1981) 817–823.
368. D. Krummen, J. Bayer, J. Ho, G. Ho, M. Smetak, P. Clopton, N. Trayanova, S. Narayan, Mechanisms of human atrial fibrillation initiation: clinical and computational studies of repolarization restitution and activation latency, Circulation: Arrhythmia and Electrophysiology 5 (6) (2012) 1149–1159.
369. F. Meister, T. Passerini, V. Mihalef, A. Tuysuzoglu, A. Maier, T. Mansi, Deep learning acceleration of total Lagrangian explicit dynamics for soft tissue mechanics, Computer Methods in Applied Mechanics and Engineering 358 (2020) 112628.
370. A. Hanson, H. Ma, Parallel Transport Approach to Curve Framing, Indiana University, Techreports-TR425, vol. 11 1995, pp. 3–7.
371. D.P. Kingma, J. Ba, Adam: a method for stochastic optimization, CoRR arXiv:1412.6980, 2014.
372. Raymond W. Ogden, Non-linear elastic deformations, Courier Corporation, 1997.
373. F. Chollet, et al., Keras, https://keras.io, 2015.
374. M. Abadi, A. Agarwal, P. Barham, E. Brevdo, Z. Chen, C. Citro, G.S. Corrado, A. Davis, J. Dean, M. Devin, S. Ghemawat, I. Goodfellow, A. Harp, G. Irving, M. Isard, Y. Jia, R. Jozefowicz, L. Kaiser, M. Kudlur, J. Levenberg, D. Mané, R. Monga, S. Moore, D. Murray, C. Olah, M. Schuster, J. Shlens, B. Steiner, I. Sutskever, K. Talwar, P. Tucker, V. Vanhoucke, V. Vasudevan, F. Viégas, O. Vinyals, P. Warden, M. Wattenberg, M. Wicke, Y. Yu, X. Zheng, TensorFlow: Large-Scale Machine Learning on Heterogeneous Systems, 2015, Software available from tensorflow.org.
375. Philippe Weinzaepfel, Jerome Revaud, Zaid Harchaoui, Cordelia Schmid, DeepFlow: Large displacement optical flow with deep matching, in: Proceedings of the IEEE International Conference on Computer Vision, 2013, pp. 1385–1392.
376. Mark Alber, Adrian Buganza Tepole, William Cannon, Suvranu De, Salvador Dura-Bernal, Krishna Garikipati, George Karniadakis, William W. Lytton, Paris Perdikaris, Linda Petzold, Ellen Kuhl, Integrating machine learning and multiscale

modeling: perspectives, challenges, and opportunities in the biological, biomedical, and behavioral sciences, arXiv preprint, arXiv:1910.01258, 2019.
377. R.M. Gulrajani, The forward and inverse problems of electrocardiography, IEEE Engineering in Medicine and Biology Magazine 17 (5) (1998) 84–101.
378. R. Modre, B. Tilg, G. Fischer, P. Wach, Noninvasive myocardial activation time imaging: a novel inverse algorithm applied to clinical ecg mapping data, IEEE Transactions on Bio-Medical Engineering 49 (10) (2002) 1153–1161.
379. O. Dössel, M.W. Krueger, F.M. Weber, C. Schilling, W.H. Schulze, G. Seemann, A framework for personalization of computational models of the human atria, in: Engineering in Medicine and Biology Society, EMBC, 2011 Annual International Conference of the IEEE, IEEE, 2011, pp. 4324–4328.
380. D. Neumann, T. Mansi, S. Grbic, I. Voigt, B. Georgescu, E. Kayvanpour, A. Amr, F. Sedaghat-Hamedani, J. Haas, H. Katus, B. Meder, J. Hornegger, A. Kamen, D. Comaniciu, Automatic image-to-model framework for patient-specific electromechanical modeling of the heart, in: IEEE International Symposium on Biomedical Imaging, 2014, pp. 935–938.
381. M. Jiang, J. Lv, C. Wang, W. Huang, L. Xia, G. Shou, A hybrid model of maximum margin clustering method and support vector regression for solving the inverse ECG problem, in: Computing in Cardiology, 2011, IEEE, 2011, pp. 457–460.
382. T. Hastie, R. Tibshirani, J. Friedman, The Elements of Statistical Learning: Data Mining, Inference, and Prediction, Springer, 2009.
383. G. Litjens, T. Kooi, B.E. Bejnordi, A.A.A. Setio, F. Ciompi, M. Ghafoorian, J.A. van der Laak, B. van Ginneken, C.I. Sánchez, A survey on deep learning in medical image analysis, arXiv:1702.05747, 2017.
384. D. Neumann, T. Mansi, L. Itu, B. Georgescu, E. Kayvanpour, F. Sedaghat-Hamedani, A. Amr, J. Haas, H. Katus, B. Meder, S. Steidl, J. Hornegger, D. Comaniciu, A self-taught artificial agent for multi-physics computational model personalization, Medical Image Analysis 34 (Dec. 2016) 52–64.
385. B. Kveton, G. Theocharous, Kernel-based reinforcement learning on representative states, in: Association for the Advancement of Artificial Intelligence, 2012, pp. 977–983.
386. D. Neumann, T. Mansi, L. Itu, B. Georgescu, E. Kayvanpour, F. Sedaghat-Hamedani, J. Haas, H. Katus, B. Meder, S. Steidl, J. Hornegger, D. Comaniciu, Vito – a generic agent for multi-physics model personalization: application to heart modeling, in: International Conference on Medical Image Computing and Computer-Assisted Intervention, in: LNCS, vol. 9350, Springer, 2015, pp. 442–449.
387. L. Itu, P. Sharma, B. Georgescu, A. Kamen, C. Suciu, D. Comaniciu, Model-based non-invasive estimation of pv loop from echocardiography, in: 36th Annual International Conference of the IEEE Engineering in Medicine and Biology Society, IEEE, 2014, pp. 6774–6777.
388. J.C. Lagarias, J.A. Reeds, M.H. Wright, P.E. Wright, Convergence properties of the Nelder-Mead simplex method in low dimensions, SIAM Journal on Optimization 9 (1) (1998) 112–147.
389. J. Cohn, Pathophysiology and clinical recognition of heart failure, in: Cardiovascular Medicine, Springer, 2007, pp. 1379–1396.
390. G. Fonarow, Risk stratification models and predictors of mortality in acute heart failure syndromes, based on United States registries, in: A. Mebazaa, M. Gheorghiade, F. Zannad, J. Parrillo (Eds.), Acute Heart Failure, Springer London, London, 2008, pp. 7–12.
391. V. Kutyifa, A. Kosztin, H.U. Klein, Y. Biton, V.K. Nagy, S.D. Solomon, S. McNitt, W. Zareba, I. Goldenberg, A. Roka, et al., Left ventricular lead location and long-term outcomes in cardiac resynchronization therapy patients, JACC: Clinical Electrophysiology 4 (11) (2018) 1410–1420.
392. R. Van Bommel, J. Bax, W. Abraham, E. Chung, L. Pires, L. Tavazzi, P. Zimetbaum, B. Gerritse, N. Kristiansen, S. Ghio, Characteristics of heart failure patients associated with good and poor response to cardiac resynchronization therapy: a prospect (predictors of response to crt) sub-analysis, European Heart Journal 30 (20) (2009) 2470–2477.
393. D. Lustgarten, S. Calame, E. Crespo, J. Calame, R. Lobel, P.S. Spector, Electrical resynchronization induced by direct his-bundle pacing, Heart Rhythm 7 (Jan 2010) 15–21.
394. D. Kass, Cardiac resynchronization therapy, Journal of Cardiovascular Electrophysiology 16 (s1) (2015) S35–S41.
395. E. Donal, V. Delgado, J. Magne, C. Bucciarelli-Ducci, C. Leclercq, B. Cosyns, M. Sitges, T. Edvardsen, E. Sade, I. Stankovic, et al., Rational and design of eurocrt: an international observational study on multi-modality imaging and cardiac resynchronization therapy, European Heart Journal-Cardiovascular Imaging 18 (10) (2017) 1120–1127.

396. E.C. Adelstein, S. Saba, Scar burden by myocardial perfusion imaging predicts echocardiographic response to cardiac resynchronization therapy in ischemic cardiomyopathy, American Heart Journal 153 (1) (2007) 105–112.
397. S. Chalil, P.W. Foley, S.A. Muyhaldeen, K.C. Patel, Z.R. Yousef, R.E. Smith, M.P. Frenneaux, F. Leyva, Late gadolinium enhancement-cardiovascular magnetic resonance as a predictor of response to cardiac resynchronization therapy in patients with ischaemic cardiomyopathy, Europace 9 (11) (2007) 1031–1037.
398. A.-C. Pouleur, D. Knappe, A.M. Shah, H. Uno, M. Bourgoun, E. Foster, S. McNitt, W.J. Hall, W. Zareba, I. Goldenberg, et al., Relationship between improvement in left ventricular dyssynchrony and contractile function and clinical outcome with cardiac resynchronization therapy: the madit-crt trial, European Heart Journal 32 (14) (2011) 1720–1729.
399. F. Leyva, R.J. Taylor, P.W. Foley, F. Umar, L.J. Mulligan, K. Patel, B. Stegemann, T. Haddad, R.E. Smith, S.K. Prasad, Left ventricular midwall fibrosis as a predictor of mortality and morbidity after cardiac resynchronization therapy in patients with nonischemic cardiomyopathy, Journal of the American College of Cardiology 60 (17) (2012) 1659–1667.
400. E.S. Chung, A.R. Leon, L. Tavazzi, J.-P. Sun, P. Nihoyannopoulos, J. Merlino, W.T. Abraham, S. Ghio, C. Leclercq, J.J. Bax, et al., Results of the predictors of response to crt (prospect) trial, Circulation (2008).
401. P. Lim, E. Donal, S. Lafitte, G. Derumeaux, G. Habib, P. Réant, S. Thivolet, N. Lellouche, R.A. Grimm, P. Gueret, Multicentre study using strain delay index for predicting response to cardiac resynchronization therapy (music study), European Journal of Heart Failure 13 (9) (2011) 984–991.
402. J. Lumens, S. Ploux, M. Strik, J. Gorcsan, H. Cochet, N. Derval, M. Strom, C. Ramanathan, P. Ritter, M. Haïssaguerre, et al., Comparative electromechanical and hemodynamic effects of left ventricular and biventricular pacing in dyssynchronous heart failure: electrical resynchronization versus left–right ventricular interaction, Journal of the American College of Cardiology 62 (25) (2013) 2395–2403.
403. N.A. Marsan, J.J. Westenberg, C. Ypenburg, R.J. van Bommel, S. Roes, V. Delgado, L.F. Tops, R.J. van der Geest, E. Boersma, A. de Roos, et al., Magnetic resonance imaging and response to cardiac resynchronization therapy: relative merits of left ventricular dyssynchrony and scar tissue, European Heart Journal 30 (19) (2009) 2360–2367.
404. K.C. Bilchick, V. Dimaano, K.C. Wu, R.H. Helm, R.G. Weiss, J.A. Lima, R.D. Berger, G.F. Tomaselli, D.A. Bluemke, H.R. Halperin, et al., Cardiac magnetic resonance assessment of dyssynchrony and myocardial scar predicts function class improvement following cardiac resynchronization therapy, JACC: Cardiovascular Imaging 1 (5) (2008) 561–568.
405. V. Delgado, R.J. van Bommel, M. Bertini, C.J.W. Borleffs, N.A. Marsan, A.C. Ng, G. Nucifora, N.R. van de Veire, C. Ypenburg, E. Boersma, et al., Relative merits of left ventricular dyssynchrony, left ventricular lead position, and myocardial scar to predict long-term survival of ischemic heart failure patients undergoing cardiac resynchronization therapy, Circulation 123 (1) (2011) 70–78.
406. C. Ypenburg, M.J. Schalij, G.B. Bleeker, P. Steendijk, E. Boersma, P. Dibbets-Schneider, M.P. Stokkel, E.E. van der Wall, J.J. Bax, Impact of viability and scar tissue on response to cardiac resynchronization therapy in ischaemic heart failure patients, European Heart Journal 28 (1) (2006) 33–41.
407. A. Crozier, B. Blazevic, P. Lamata, G. Plank, M. Ginks, S. Duckett, M. Sohal, A. Shetty, C.A. Rinaldi, R. Razavi, et al., The relative role of patient physiology and device optimisation in cardiac resynchronisation therapy: a computational modelling study, Journal of Molecular and Cellular Cardiology 96 (2016) 93–100.
408. P.R. Huntjens, J. Walmsley, S. Ploux, P. Bordachar, F.W. Prinzen, T. Delhaas, J. Lumens, Influence of left ventricular lead position relative to scar location on response to cardiac resynchronization therapy: a model study, Europace 16 (suppl 4) (2014), iv62–iv68.
409. E. Willemen, R. Schreurs, P.R. Huntjens, M. Strik, G. Plank, E. Vigmond, J. Walmsley, K. Vernooy, T. Delhaas, F.W. Prinzen, et al., The left and right ventricles respond differently to variation of pacing delays in cardiac resynchronization therapy: a combined experimental-computational approach, Frontiers in Physiology 10 (2019) 17.
410. A.W. Lee, A. Crozier, E.R. Hyde, P. Lamata, M. Truong, M. Sohal, T. Jackson, J.M. Behar, S. Claridge, A. Shetty, et al., Biophysical modeling to determine the optimization of left ventricular pacing site and av/vv delays in the acute and chronic phase of cardiac resynchronization therapy, Journal of Cardiovascular Electrophysiology 28 (2) (2017) 208–215.

411. S.A. Niederer, A. Shetty, G. Plank, J. Bostock, R. Razavi, N. Smith, C. Rinaldi, Biophysical modeling to simulate the response to multisite left ventricular stimulation using a quadripolar pacing lead, Pacing and Clinical Electrophysiology 35 (2) (2012) 204–214.
412. C.M. Costa, A. Neic, G. Plank, E. Kerfoot, B. Porter, B. Sieniewicz, J. Gould, B. Sidhu, Z. Chen, C.A. Rinaldi, et al., Personalized Modeling Pipeline for Cardiac Electrophysiology Simulations of Cardiac Resynchronization Therapy in Infarct Patients, 2018 Computing in Cardiology Conference (CinC), vol. 45, IEEE, 2018, pp. 1–4.
413. A.F. Members, J.J. McMurray, S. Adamopoulos, S.D. Anker, A. Auricchio, M. Böhm, K. Dickstein, V. Falk, G. Filippatos, C. Fonseca, et al., Esc guidelines for the diagnosis and treatment of acute and chronic heart failure 2012: the task force for the diagnosis and treatment of acute and chronic heart failure 2012 of the European society of cardiology. developed in collaboration with the heart failure association (hfa) of the esc, European Journal of Heart Failure 14 (8) (2012) 803–869.
414. C. Ferencz, J.D. Rubin, R.J. Mccarter, J.I. Brenner, C.A. Neill, L.W. Perry, S.I. Hepner, J.W. Downing, Congenital heart disease: prevalence at livebirth: the Baltimore-Washington infant study, American Journal of Epidemiology 121 (1) (1985) 31–36.
415. M. Cohen, V. Fuster, P.M. Steele, D. Driscoll, D.C. Mcgoon, Coarctation of the aorta. long-term follow-up and prediction of outcome after surgical correction, Circulation 80 (4) (1989) 840–845.
416. G. Webb, Treatment of coarctation and late complications in the adult, in: Seminars in Thoracic and Cardiovascular Surgery, vol. 17, Elsevier, 2005, pp. 139–142.
417. R. Puranik, V.T. Tsang, S. Puranik, R. Jones, S. Cullen, P. Bonhoeffer, M.L. Hughes, A.M. Taylor, Late magnetic resonance surveillance of repaired coarctation of the aorta, European Journal of Cardio-Thoracic Surgery 36 (1) (2009) 91–95.
418. W.E. Hellenbrand, H.D. Allen, R.J. Golinko, D.J. Hagler, W. Lutin, J. Kan, Balloon angioplasty for aortic recoarctation: results of valvuloplasty and angioplasty of congenital anomalies registry, American Journal of Cardiology 65 (11) (1990) 793–797.
419. C. Ovaert, L. Benson, D. Nykanen, R. Freedom, Transcatheter treatment of coarctation of the aorta: a review, Pediatric Cardiology 19 (1) (1998) 27–44.
420. D. Tanous, L.N. Benson, E.M. Horlick, Coarctation of the aorta: evaluation and management, Current Opinion in Cardiology 24 (6) (2009) 509–515.
421. E. Rosenthal, Stent Implantation for Aortic Coarctation: The Treatment of Choice in Adults?, 2001.
422. B.L. Seifert, K. DesRochers, M. Ta, G. Giraud, M. Zarandi, M. Gharib, D.J. Sahn, Accuracy of Doppler methods for estimating peak-to-peak and peak instantaneous gradients across coarctation of the aorta: an in vitro study, Journal of the American Society of Echocardiography 12 (9) (1999) 744–753.
423. G.-Z. Yang, P.J. Kilner, N.B. Wood, S.R. Underwood, D.N. Firmin, Computation of flow pressure fields from magnetic resonance velocity mapping, Magnetic Resonance in Medicine 36 (4) (1996) 520–526.
424. T. Ebbers, L. Wigström, A.F. Bolger, J. Engvall, M. Karlsson, Estimation of relative cardiovascular pressures using time-resolved three-dimensional phase contrast mri, Magnetic Resonance in Medicine 45 (5) (2001) 872–879.
425. D. Firmin, P. Gatehouse, J. Konrad, G. Yang, P. Kilner, D. Longmore, Rapid 7-dimensional imaging of pulsatile flow, in: Computers in Cardiology 1993, Proceedings, IEEE, 1993, pp. 353–356.
426. L. Wigström, L. Sjöqvist, B. Wranne, Temporally resolved 3d phase-contrast imaging, Magnetic Resonance in Medicine 36 (5) (1996) 800–803.
427. Z. Qian, X. Zhou, M. Vannan, 4d flow imaging in aortic disease, in: Advances in Treatments for Aortic Valve and Root Diseases, Springer, 2018, pp. 137–143.
428. J.J. Hom, K. Ordovas, G.P. Reddy, Velocity-encoded cine mr imaging in aortic coarctation: functional assessment of hemodynamic events, Radiographics 28 (2) (2008) 407–416.
429. L.J. Olivieri, D.A. de Zélicourt, C.M. Haggerty, K. Ratnayaka, R.R. Cross, A.P. Yoganathan, Hemodynamic modeling of surgically repaired coarctation of the aorta, Cardiovascular Engineering and Technology 2 (4) (2011) 288–295.
430. A. Menon, D.C. Wendell, H. Wang, T.J. Eddinger, J.M. Toth, R.J. Dholakia, P.M. Larsen, E.S. Jensen, J.F. LaDisa Jr, A coupled experimental and computational approach to quantify deleterious hemodynamics, vascular alterations, and mechanisms of long-term morbidity in response to aortic coarctation, Journal of Pharmacological and Toxicological Methods 65 (1) (2012) 18–28.
431. J. Lantz, J. Renner, M. Karlsson, Wall shear stress in a subject specific human aorta—influence of fluid-structure interaction, International Journal of Applied Mechanics 3 (04) (2011) 759–778.
432. J.S. Coogan, J.D. Humphrey, C.A. Figueroa, Computational simulations of hemodynamic

changes within thoracic, coronary, and cerebral arteries following early wall remodeling in response to distal aortic coarctation, Biomechanics and Modeling in Mechanobiology 12 (1) (2013) 79–93.

433. J.F. LaDisa, C.A. Figueroa, I.E. Vignon-Clementel, H.J. Kim, N. Xiao, L.M. Ellwein, F.P. Chan, J.A. Feinstein, C.A. Taylor, Computational simulations for aortic coarctation: representative results from a sampling of patients, Journal of Biomechanical Engineering 133 (9) (2011) 091008.

434. D. Gallo, G. De Santis, F. Negri, D. Tresoldi, R. Ponzini, D. Massai, M.A. Deriu, P. Segers, B. Verhegghe, G. Rizzo, et al., On the use of in vivo measured flow rates as boundary conditions for image-based hemodynamic models of the human aorta: implications for indicators of abnormal flow, Annals of Biomedical Engineering 40 (3) (2012) 729–741.

435. P. Reymond, Y. Bohraus, F. Perren, F. Lazeyras, N. Stergiopulos, Validation of a patient-specific one-dimensional model of the systemic arterial tree, American Journal of Physiology. Heart and Circulatory Physiology 301 (3) (2011) H1173–H1182.

436. L. Itu, P. Sharma, K. Ralovich, V. Mihalef, R. Ionasec, A. Everett, R. Ringel, A. Kamen, D. Comaniciu, Non-invasive hemodynamic assessment of aortic coarctation: validation with in vivo measurements, Annals of Biomedical Engineering 41 (4) (2013) 669–681.

437. L. Goubergrits, E. Riesenkampff, P. Yevtushenko, J. Schaller, U. Kertzscher, F. Berger, T. Kuehne, Is mri-based cfd able to improve clinical treatment of coarctations of aorta?, Annals of Biomedical Engineering 43 (1) (2015) 168–176.

438. D.F. Young, F.Y. Tsai, Flow characteristics in models of arterial stenoses—i. Steady flow, Journal of Biomechanics 6 (4) (1973) 395–402.

439. L. Liang, M. Liu, C. Martin, W. Sun, A deep learning approach to estimate stress distribution: a fast and accurate surrogate of finite-element analysis, Journal of the Royal Society Interface 15 (138) (2018) 20170844.

440. L. Liang, M. Liu, C. Martin, J.A. Elefteriades, W. Sun, A machine learning approach to investigate the relationship between shape features and numerically predicted risk of ascending aortic aneurysm, Biomechanics and Modeling in Mechanobiology 16 (5) (2017) 1519–1533.

441. O. Hennigh, Lat-net: compressing lattice Boltzmann flow simulations using deep neural networks, arXiv preprint, arXiv:1705.09036, 2017.

442. J. Tompson, K. Schlachter, P. Sprechmann, K. Perlin, Accelerating Eulerian fluid simulation with convolutional networks, arXiv preprint, arXiv:1607.03597, 2016.

443. E. Jones, T. Oliphant, P. Peterson, et al., SciPy: Open Source Scientific Tools for Python, 2001, Online; accessed 2019/08/30.

444. S. Chikatamarla, S. Ansumali, I. Karlin, Entropic lattice Boltzmann models for hydrodynamics in three dimensions, Physical Review Letters 97 (1) (2006) 010201.

445. W. Xian, A. Takayuki, Multi-gpu performance of incompressible flow computation by lattice Boltzmann method on gpu cluster, Parallel Computing 37 (9) (2011) 521–535.

446. C. Obrecht, F. Kuznik, B. Tourancheau, J.-J. Roux, Multi-gpu implementation of the lattice Boltzmann method, Computers & Mathematics with Applications 65 (2) (2013) 252–261.

447. J. Tölke, Implementation of a lattice Boltzmann kernel using the compute unified device architecture developed by nvidia, Computing and Visualization in Science 13 (1) (2010) 29–39.

448. C. Nita, L.M. Itu, C. Suciu, Gpu accelerated blood flow computation using the lattice Boltzmann method, in: High Performance Extreme Computing Conference (HPEC), 2013 IEEE, IEEE, 2013, pp. 1–6.

449. X. Liu, Y. Fan, X. Deng, Effect of spiral flow on the transport of oxygen in the aorta: a numerical study, Annals of Biomedical Engineering 38 (3) (2010) 917–926.

450. Y. Huo, M. Svendsen, J.S. Choy, Z.-D. Zhang, G.S. Kassab, A validated predictive model of coronary fractional flow reserve, Journal of the Royal Society Interface (2011), rsif20110605.

451. M. Abadi, A. Agarwal, P. Barham, E. Brevdo, Z. Chen, C. Citro, G.S. Corrado, A. Davis, J. Dean, M. Devin, et al., Tensorflow: large-scale machine learning on heterogeneous distributed systems, arXiv preprint, arXiv:1603.04467, 2016.

452. D.P. Kingma, J. Ba, Adam: a method for stochastic optimization, arXiv preprint, arXiv:1412.6980, 2014.

453. V. Mihalef, L. Itu, T. Mansi, P. Sharma, Lumped Parameter Whole Body Circulation Modelling, Springer International Publishing, Cham, 2017.

454. A. Vizitiu, C.I. Nita, A. Puiu, C. Suciu, L.M. Itu, Privacy-preserving artificial intelligence: application to precision medicine, in: 41st International Engineering in Medicine and Biology Conference, 2019.

455. D. Mann, D. Zipes, P. Libby, R. Bonow, Braunwald's Heart Disease e-Book: A Textbook of

Cardiovascular Medicine, Elsevier Health Sciences, 2014.

456. D. Burkhoff, Pressure-volume loops in clinical research: a contemporary view, Journal of the American College of Cardiology 62 (2013) 1173–1176.

457. D.M. Spevack, J. Karl, N. Yedlapati, Y. Goldberg, M.J. Garcia, Echocardiographic left ventricular end-diastolic pressure volume loop estimate predicts survival in congestive heart failure, Journal of Cardiac Failure 19 (2013) 251–259.

Index

9780128175941